Study Guide to Accompany
The World of Biology

THIRD EDITION

by Davis and Solomon

David Fromson, Ph.D.
Professor of Biology
California State University—Fullerton

Kenneth Goodhue-McWilliams, Ph.D.
Professor of Biology
California State University—Fullerton

Ina Katz, Ph.D.
Coordinator, Learning Assistance and Resource Center
California State University—Fullerton

SAUNDERS COLLEGE PUBLISHING
Philadelphia New York Chicago
San Francisco Montreal Toronto
London Sydney Tokyo Mexico City
Rio de Janeiro Madrid

Address orders to:
383 Madison Avenue
New York, NY 10017

Address editorial correspondence to:
West Washington Square
Philadelphia, PA 19105

Study Guide to accompany THE WORLD OF BIOLOGY

ISBN # 0-03-059998-9

Printed in the United States of America.

5678 066 987654321

CBS COLLEGE PUBLISHING
Saunders College Publishing
Holt, Rinehart and Winston
The Dryden Press

PREFACE

We wrote this Study Guide which accompanies The World of Biology by Davis & Solomon to help develop study skills for students taking their first college level biology course. The various learning aids and study strategies covered in this Study Guide have been successfully used in our own science courses. Many study aids, including this one, are based on the principle that students learn by doing. This study guide provides students an opportunity to learn how to study biology and consequently how to learn from the text, The World of Biology by Davis & Solomon.

FORMAT. The outline below indicates how each chapter in the guide is structured.

CHAPTER 1

LIFE, SCIENCE AND SOCIETY

BEFORE READING ACTIVITIES

- chapter outline preview and questions
- vocabulary preview

DURING READING ACTIVITIES

- vocabulary
- marking the text
- reading the figures

AFTER READING ACTIVITIES

- review of textbook marking
- vocabulary review
- sample test questions
- answers to questions

THE LANGUAGE OF BIOLOGY

- selected terms on facing page with definitions on back page

The study skills we present will help you use The World of Biology efficiently and effectively. By acquiring and developing the basic study skills presented here we trust you will find your biology course a rewarding and successful experience.

David Fromson
Kenneth Goodhue-McWilliams
Ina Katz

February, 1986

CONTENTS

PART VI ANIMAL STRUCTURE AND FUNCTION

PART VII EVOLUTION, BEHAVIOR, AND ECOLOGY

CHAPTER 1

LIFE, SCIENCE, AND SOCIETY

BEFORE READING ACTIVITIES

I. Before you begin to read the chapter, look over the outline below and answer the questions which follow it. Be sure you write your answers down so you can check them later.

I. What is life?
   A. Cellular structure
   B. Growth
   C. Metabolism
   D. Homeostasis
   E. Movement
   F. Responsiveness
   G. Reproduction
   H. Adaptation
II. Life's Organization
   A. The organization of the organism
   B. Econological organization
III. Society and the ecosphere
IV. The variety of organisms
V. How biology is studied: scientific method
   A. Science in action
      1. Formulating a hypothesis
      2. Setting up an experiment
   B. How hypotheses become theories or principles
   C. The ethics of science

Focus on the evolutionary perspective

Focus on recognizing a problem: the prepared mind

Questions - Remember to refer to the outline above.

1. List the characteristics of living systems. (Hint: See I above.) What are they?

   1. ________________ 5. ________________
   2. ________________ 6. ________________
   3. ________________ 7. ________________
   4. ________________ 8. ________________

2. Write down your best guess for the definition of each of these characteristics.

   1. ________________ 5. ________________
   2. ________________ 6. ________________
   3. ________________ 7. ________________
   4. ________________ 8. ________________

3. What are the two major categories through which biologists examine the organization of life?
   1. ____
   2. ____

4. What is the ecosphere? Make a guess!

5. What do you think is meant by "variety of organisms"?

6. What are the important parts of the scientific method?

II. Take the quiz on page 24-25 of the test to find out how much you already know. The summary of pages 23-24 may be helpful.

III. Sort the vocabulary cards which follow into the categories of the outline where you think they will be discussed. To sort the vocabulary cards, first tear out the sheet. Then separate the words into individual cards. Read the words on the front and the definitions on the back. Finally, decide into which Roman numeral of the chapter outline they are most likely to fit.

DURING READING ACTIVITIES

I. Vocabulary - Read the chapter in major chapter sections (from one Roman numeral to the next). As you read, check to see if you have vocabulary cards for each new term you find. Be sure to circle the term in the textbook and then find the vocabulary card that defines that term.

II. Marking the text. Underline or highlight the text as you read. This is important because, if you do it carefully, you should not have to read the whole chapter again to prepare for examinations. Use the following procedure to begin marking your text:

1. Read a whole paragraph through completely WITHOUT making a single mark on the page. AFTER you have completed the paragraph, ask yourself, "What is important about this paragraph?" Now, reread the paragraph and mark accordingly.

2. Read each of the paragraphs in a major section and use the same marking procedure. When you have completed a whole major section, go back over your highlighting and see if it makes sense to you now that you have had some time to forget what you read earlier.

III. Reading Figures. Be sure to study the Figures in the textbook. Use the following procedure:

1. Look over the non-verbal portion of the figure. See if you can explain what it means WITHOUT reading the verbal explanation.

2. Read the verbal explanation of the figure. Compare your explanation with the one given in the text: How close did you get to the author's explanation?

3. Try to decide how the figure fits into the information given in the textbook.

For an example, look at the Figure 1-5, page 7.

1. What characteristics of living systems does it relate to?

2. Is this the only mechanism for this characteristic or is it an example of one particular way this mechanism works?

3. What type of living creatures have this mechanism?

AFTER READING ACTIVITIES

I. Reviewing your textbook marking.

When you finish reading for this time period, reread what you have underlined in the textbook. If there are portions which do not make sense, reread the paragraph and correct your underlining.

II. Review the vocabulary.

As you find a circled term, check to see if you know the definition. Sort the cards into two piles: words you know and words you need to learn. Check to see if you have used all the vocabulary cards. If you have some left over, find the terms in the text. Check to see if you know the definitions. If these are familiar terms, put them in the pile with words you know; if they are unfamiliar, put them in the pile of words you need to learn.

III. Answer the following questions:

1. The characteristics which enable an organism to better survive in a particular environment are the ____________________________ characteristics.

2. Organisms maintain a constant, stable internal environment. This is _______________.

3. The ________________ reactions in an organism which liberate energy as well as the complex reactions that form energy and form new tissues are the organisms's _______________.

4. _______________ is the study of life.

5. A biological _______________ consists of various interacting populations of different species occupying the same area.

6. Cells of higher organisms are divided into a ______________ and a _______________. The _______________ is the organelle which contains the hereditary material DNA.

7. __________________ are arrangements of cells into functional units.

8. The study of the interactions of organisms in a community is called _______________.

9. The smallest amount of a chemical element which still retains the properties and characteristics of that element is a(n) ______________.

10. A group of organisms of a particular species is called a ______________.

11. ______________ is the process in which food is produced from raw materials utilizing sunlight as the energy source.

12. Organisms which are dependent on the producers for food, oxygen and energy are the ______________.

13. A tentitive explanation for a particular phenomenon is the ______________.

14. The ______________ and ______________ are the decomposers which break down wastes and the bodies of dead organisms.

15. When a tentative explanation is supported by a large body of experiemental data and many observations, this explanation becomes a ______________.

16. A ______________ is a combination of atoms.

17. Cells associate in a specific way to form a ______________.

18. ______________ ______________ is the process in which a single parent organism buds or divides to give rise to offspring.

19. When the genetic or hereditary material is changed, a ______________ has occurred.

20. Organisms are sensitive to change and thus said to be ______________.

21. Match the following terms:

1.	Fungi	(a)	Plants
2.	Animalia	(b)	Cyanobacteria, blue-green algae, bacteria
3.	Protista	(c)	Protozoa, some algae
4.	Monera	(d)	Animals
5.	Plantae	(e)	Fungi and slime molds

22. What are the characteristics of living systems or organisms?

23. Distinguish between asexual and sexual reproduction.

24. If an organism is adapted to its environment, what physiological processes is it likely to carry out?

25. For each of the paired items below, indicate which is higher in terms of BIOLOGICAL ORGANIZATION.

A. Molecule - Cell
B. Atom - Organelle
C. Organelle - Tissue
D. Organ - Organism

26. What is the <u>difference</u> between a population and a community of organisms?

27. What would be the consequences of a sudden loss of all the decomposers on earth?

28. Set up an experiment to test the hypothesis that using a commercial plant fertilizer and water will produce larger, healthier plants more quickly than water alone.

29. What are the characteristics of a good hypothesis?

30. What are the three types of organisms of an ecosystem?

31. What are the six steps which constitute the scientific method?

32. What is the basic unit of the living organism?

Chapter I - Answers

1. Adaptive
2. Homostasis
3. Chemical, metabolism
4. Biology
5. Community
6. Nucleus, cytoplasm, nucleus
7. Organs
8. Ecology
9. Atom
10. Population
11. Photosynthesis
12. Consumers
13. Hypothesis
14. Bacteria, fungi
15. Theory
16. Molecule
17. Tissue
18. Asexual reproduction
19. Mutation
20. Irritable
21.
    1. e
    2. d
    3. c
    4. b
    5. a

22.
    A. Growth
    B. Self-regulated metabolism
    C. Movement
    D. Response
    E. Reproduction
    F. Adaptation to environment change

23. Work this one out and check the text.

24. Growth, reproduction

25.
    A. Cell
    B. Organelle
    C. Tissue
    D. Organism

26. A community consists of several species of organisms while a population consists of many individuals of the same species.

27. Vast accumulation of wastes, interruption of cycles.

28. Do this one on your own.

29.
    A. Consistent with facts
    B. Testable
    C. Similer than other possible hypotheses

30. Producers, consumers, decomposers.

31.
    A. Recognize and state the problem
    B. Collect data (information) on the problem.
    C. Formulate a testable hypothesis.
    D. Make a prediction on the hypothesis.

E. Make observations or design and perform experiments to test hypothesis.
F. Formulate a conclusion

32. The cell.

BIOLOGY

LIVING

GROWTH

METABOLISM

HOMEOSTASIS

IRRITABLE

REPRODUCTION

ASEXUAL PRODUCTION

SEXUAL REPRODUCTION

ADAPTATION

MUTATION

ATOM

SYSTEM OR ORGANISM.

THE STUDY OF LIFE - ALL LIVING THINGS ON EARTH.

ALL THE CHEMICAL REACTIONS WHICH TAKE PLACE IN AN ORGANISM, INCLUDING THOSE WHICH LIBERATE ENERGY AND THE COMPLEX REACTIONS TO FORM ENERGY. ALSO MAINTAIN THE CELL, TISSUES AND ORGANISMS.

RAW MATERIAL FROM THE ENVIRONMENT AND FORMING THE ORGANISM'S (CELL'S) OWN TYPE OF SUBSTANCES.

SENSITIVE TO CHANGE.

TENDENCY TO MAINTAIN A CONSTANT INTERNAL ENVIRONMENT.

SINGLE PARENT, BUDS, OR DIVIDES TO GIVE RISE TO OFFSPRING.

CHARACTERISTICS THAT ENABLE AN ORGANISM TO SURVIVE IN A SPECIFIC ENVIRONMENT.

FUSION OF SPECIAL CELLS - AN EGG AND A SPERM TO FORM A FERTILIZED EGG.

SMALLEST AMOUNT OF CHEMICAL ELEMENT WHICH IS CHARACTERISTIC OF THAT ELEMENT.

CHANGE IN THE HEREDITORY MATERIAL OF THE CELL.

MOLECULES

CELLULAR LEVEL

CELLS

ORGANS

ORGAN SYSTEM

ORGANISM

POPULATION

COMMUNITY

ECOSYSTEM

ECOLOGY

PRODUCERS

PHOTOSYNTHESIS

ORGANIZATION OF MOLECULES - DIVIDED INTO NUCLEUS AND CYTOPLASM. NUCLEUS ORGANELLE - CONTAINS THE GENES (HEREDITORY MATERIAL). CYTOPLASM - COMPLEX CHEMICAL MIXTURE AND SUBCELLULAR ORGANELLES.

COMBINATION OF ATOMS.

ARRANGEMENT OF CELLS INTO FUNCTIONAL UNITS.

ASSOCIATE TO FORM TISSUES.

COMBINATION OF ORGAN SYSTEMS.

BIOLOGICAL FUNCTION BY COORDINATED GROUP OF TISSUES AND ORGANS.

VARIOUS INTERACTING POPULATIONS OF DIFFERENT SPECIES OCCUPYING SAME AREA - HUNDREDS OR THOUSANDS OF DIFFERENT TYPES OR ORGANISMS.

GROUP OF ORGANISMS OF A PARTICULAR SPECIES.

STUDY OF HOW ORGANISMS OF COMMUNITY INTERACT WITH ONE ANOTHER.

COMMUNITY PLUS NONLIVING ENVIRONMENT - SELF SUFFICIENT.

PRODUCTION OF FOOD FROM RAW MATERIALS USING SUNLIGHT AS AN ENERGY SOURCE.

ORGANISMS THAT PRODUCE THEIR OWN FOOD FROM RAW MATERIALS.

CONSUMERS

CELLULAR RESPIRATION

DECOMPOSERS

GENUS AND SPECIES

MONERA

PROTISTA

FUNGI

PLANTAE

ANIMALIA

HYPOTHESIS

THEORY

STORED ENERGY MADE AVAILABLE.

PROTOZOA - SOME ALGAE.

PLANTS

TENTATIVE EXPLANATION.

HYPOTHESIS THAT IS SUPPORTED BY LARGE BODY OF EXPERIMENTAL DATA AND OBSERVATIONS.

DEPENDENT ON THE PRODUCERS FOR FOOD, OXYGEN AND ENERGY. BREAKDOWN FOOD MOLECULES ORIGINALLY PRODUCED BY PHOTOSYNTHESIS - THIS IS CELLULAR RESPIRATION.

BACTERIA AND FUNGI - BREAK-DOWN WASTES AND BODIES OF DEAD ORGANISMS - YIELDS REUSABLE CONSTITUENTS.

BACTERIA AND BLUE-GREEN ALGAE.

FUNGI AND SLIME MOLDS - DECOMPOSERS.

ANIMALS.

CHAPTER 2

THE CHEMISTRY OF LIFE: ATOMS AND MOLECULES

BEFORE READING ACTIVITIES

I. Before you begin to read the chapter, look over the outline below and answer the questions which follow it. Be sure you write your answers so you can check them after you read the chapter. Remember, you are NOT supposed to know all the answers before you read the chapter.

I. Chemical elements

II. The storm
- A. Atomic structure
- B. Electron configuration

III. Chemical compounds
- A. What are formulas?
- B. Chemical equations

IV. How atoms combine: chemical bonds
- A. Covalent bonds
- B. Ionic bonds
- C. Hydrogen bonds

V. Oxidation and reduction

VI. Inorganic compounds
- A. Water
    1. Polar properties
    2. Properties as a solvent
    3. Cohesive and adhesive forces
    4. Temperature stabilization
- B. Acids and bases
- C. Salts
- D. Buffers

Focus on isotopes

Questions

1. What are chemical elements? Can you name some chemical elements?

2. What two characteristics of atoms are important? How do these characteristics differ?

3. How do you think an atom is different from a chemical compound?

4. What do you think chemical bonding is? List three types of chemical bonds.

5. What groups of inorganic compounds will you read about? What do you already know about them?

6. How does the section titled "Focus on Isotopes" fit into the rest of the chapter?

II. Read the summary on page 42 of the textbook. Take the quiz on pages 42-43 to find out how much you already know.

III. Use the vocabulary page as follows:

1. As you separate the sheet into individual cards, read the terms on the front and the definitions on the back.

2. Sort the terms into groups which you think will fit into the major sections of the chapter (I through VI).

3. How many of the terms did you already know?

DURING READING ACTIVITIES

I. Read the chapter in major sections (from one Roman Numeral to the next).

II. Vocabulary

1. As you read, check to see if you have vocabulary cards for each new term you find.

2. Circle the term in the textbook and find the vocabulary card that defines it.

3. Try to define the term BEFORE you look at the definition.

4. If you find terms which are new to you and not on the vocabulary cards, use the blank cards to add these terms to your set.

III. Marking the text

1. After you read each paragraph, mark the most important information by highlighting or underlining. Remember to read the whole paragraph first and then go back and mark ONLY the most important information. The example below is from page 31 of the textbook.

A. Only the key words needed to make sense of the meaning are marked.

B. The last portion of the paragraph is marked "example" to remind you that if you understand the concept (in this case of what a chemical formula is) then you do NOT need to review the example.

What Are Formulas?

A chemical formula is a shorthand method for describing the chemical composition of a molecule. Chemical symbols are used to indicate the types of atoms in the molecule, and subscript numbers are used to indicate the number of each type of atom present. The chemical formula for molecular oxygen, O_2, tells us that this molecule consists of two atoms of oxygen. The chemical formula for water, H_2O, indicates that each molecule consists of two atoms of hydrogen and one atom of oxygen. (Note that when a single atom of one type is present, it is not necessary to write 1; one does NOT write H_2O_1.)

2. After you have read a complete section (from one Roman Numeral to the next) reread your marking to be sure it makes sense to you.

3. Write notes in the margin explaining the important concepts in your OWN WORDS. Also list any groups of things that you believe you will need to remember. For example, on page 27 you should have listed the 6 elements that make up about 98 percent of an organism: oxygen, carbon, hydrogen, nitrogen, calcium, and phosphorus.

IV. Reading Figures

1. Turn to Figure 2-3 on page 33.

2. Look at the non-verbal portion of the figure. Explain what it means to yourself. For example, "one atom of hydrogen added to a second atom of hydrogen form a molecule of hydrogen. The notation for this is H-H."

3. How does this figure clarify information already presented in the text? Will you need to be able to make figures like this one?

AFTER READING ACTIVITIES

I. Review your textbook marking. Be sure that you understand what you marked and why it is important. If you think you left out something important, mark it now.

II. Review the vocabulary.

1. Go through each term you circled and repeat the definition. Sort the vocabulary cards into the words you know and the words you need to learn.

2. Begin studying the terms you need to know in groups of 5-7 words. Try to practice each new group several times each day. Don't forget to review the previous day's words before you begin each day's new word.

III. Answer the questions which follow:

At the end of this chapter you will find vocabulary cards for most of the vocabulary terms introduced in this chapter. Use these cards to learn the meaning of these terms and then answer the questions below.

1. ________________ are atoms which have 5, 6, or 7 outer shell electrons and these atoms tend to gain electrons from other atoms. They are ______________ charged.

2. The force of electrical attraction between two oppositely charged ions is called a(n) ______________ bond. These bonds can form a compound when the bonds form between an electron donor and an ____________ acceptor.

3. The loss of electrons is ______________.

4. When a substance gains electrons the substance is ___________.

5. The bond formed when a hydrogen atom is attracted to two electro-negative atoms such as oxygen and nitrogen is called a ______________ bond.

6. An atom which has lost its valence electrons is a ______________, and it is a ______________ ion.

7. Substances which cannot be broken down into simplier substances are ______________.

8. Chemists usually use abbreviations for the chemical ______________. These abbreviations are usually one or two letters and are called ______________ ______________.

9. A(n) ______________ is the smallest subdivision of an element which retains the properties of that element.

10. There are three major subatomic particles which make up an atom. These are the ______________ which have a positive electric charge, the ______________ which have no charge, and the smallest particles the ______________ which have a ______________ charge.

11. Atoms have a fixed number of protons in the atomic nucleus of a particular element. This number is the ______________ ______________ and is specific for each of the elements.

12. The total number of protons and neutrons in an atomic nucleus is the atomic ______________ ______________.

13. A combination of two or more atoms is a ______________.

14. A combination of two or more different elements in a specific ratio is a ______________ ______________.

15. The atoms which have their outer electron orbitals filled and are therefore chemically <u>reactive or nonreactive</u> (choose one) are called the ______________ ______________.

16. When the outermost shell of an atom contains less than eight electrons, it will lose, gain or share these ______________ electrons to get eight electrons in its outer shell.

17. The ______________ scale is used to indicate the degree of acidity or alkalinity of a substance. The ______________ scale ranges from ______________ to ______________.

18. A ______________ is a proton acceptor and an ______________ is a proton donor.

19. Substances which can resist pH changes when acids or bases are added are ______________.

20. An electron _______________ is the energy level of the electron is around an atomic nucleus. The _______________ energy level is closest to the atomic nucleus.

21. Bohr models are simple diagrams which show the distribution of _______________ around the atomic _______________.

22. The attractive force (general) that holds two atoms together is a _______________ _______________.

23. When two pairs of electron are shared by two atoms, this is called a _______________ _______________ bond.

24. List the elements which comprise nearly all of the mass of a living organism.

25. Is a molecule of oxygen gas the same as a molecule of liquid oxygen?

26. Indicate the ATOMIC NUMBER and ATOMIC MASS NUMBER for calcium, carbon, hydrogen and oxygen. Also write the abbreviations for each of these elements.

27. Why is it important to know how electrons are arranged around the atomic nucleus?

28. How many electron pairs are shared in a <u>triple</u> covalent bond?

29. Using Table 2-2, list three anions and three cations.

30. Using Table 2-2, list three compounds which have ionic bonds.

31. What are the three types of chemical bonds commonly found in biologically important molecules and compounds?

32. List the properties of water that help it to allow life on earth to exist.

33. What property of water makes it an ideal liquid for use in an automobile radiator?

34. What makes hydrochloric acid a stronger acid than acetic acid (vinegar)?

35. Is bath soap acidic or basic?

36. Is lemonade acidic or basic?

Chapter 2 - Answers

1. Anions, negatively
2. Ionic, electron
3. Oxidation
4. Reduced
5. Hydrogen
6. Cation, positive
7. Elements
8. Elements, chemical symbols
9. Atoms
10. Protons, neutrons, electrons, negative
11. Atomic number
12. Mass number
13. Molecule
14. Chemical compound
15. Nonreactive, noble gases
16. Valence
17. PH, pH, 0, 14
18. Base, acid
19. Buffers
20. Shell or level, lowest
21. Electrons, nucleus
22. Chemical bond
23. Double covalent
24. Oxygen, carbon, hydrogen, nitrogen, calcium, and phosphorus
25. Yes
26. Ca, ${}^{12}_{6}C$, ${}^{1}_{1}H$, and ${}^{16}_{8}O$
27. This information allows chemists to predict how particular atoms will combine
28. Three
29. See Table 2-2
30. See Table 2-2
31. 1. Covalent bonds
    2. Hydrogen bonds
    3. Ionic bonds
32. Polar properties
    Solvent properties
    Cohesive and adhesive properties
    Temperature stabilization properties
33. Temperature stabilization
34. Hydrochloric acid looses all its protons when in water
35. Basic
36. Acidic

ELEMENTS

TRACE ELEMENTS

CHEMICAL CYMBOLS

ATOM

SUBATOMIC PARTICLES

PROTONS

NEUTRONS

ELECTRONS

ATOMIC NUCLEUS

ATOMIC NUMBER

ATOMIC MASS NUMBER

ELECTRON CONFIGURATION

ELEMENTS PRESENT IN LIVING ORGANISMS IN MINUTE QUANTITIES.

SUBSTANCES WHICH CANNOT BE BROKEN DOWN INTO SIMPLER SUBSTANCES. 92 NATURAL ELEMENTS, 17 MAN-MADE ELEMENTS. O,C,H.N,CAPCHON

SMALLEST SUBDIVISION OF AN ELEMENT WHICH RETAINS PROPERTIES OF THAT ELEMENT.

ABBREVIATIONS USED BY CHEMISTS FOR ELEMENTS--USUALLY FIRST OR FIRST AND SECOND LETTER.

POSITIVE ELECTRIC CHARGE.

PARTICLES WHICH MAKE UP AN ATOM.

PROTON NEGATIVE CHARGE - SMALL MASS 1/1800 MASS OF PROTON.

UNCHARGED PARTICLES - SAME MASS AS PROTON.

THE FIXED NUMBER OF PROTONS IN THE ATOMIC NUCLEUS OF A PARTICULAR ELEMENT.

WHERE NEUTRONS AND PROTONS ARE CONCENTRATED.

ARRANGEMENT OF ELECTRONS AROUND THE ATOMIC NUCLEUS.

TOTAL NUMBER OF PROTONS PLUS NEUTRONS IN ATOMIC MASS NUMBER INDICATED BY SUPERSCRIPT TO LEFT OF SYMBOL AND THE ATOMIC NUMBER IS WRITTEN AS A SUBSCRIPT TO THE LEFT.

ELECTRONIC LEVEL OR ELECTRON SHELL

BOHR MODELS

ORBITALS

ELECTRON CLOUD

MOLECULE

CHEMICAL COMPOUND

CHEMICAL FORMULA

STRUCTURAL FORMULA

CHEMICAL EQUATION

NOBE GASES

VALENCE ELECTRONS

CHEMICAL BOND

SIMPLE DIAGRAMS SHOWING ELECTRONS AROUND THE ATOMIC NUCLEUS.

ENERGY LEVELS OF THE ELECTRONS AROUND THE ATOMIC NUCLEUS. LOWEST ENERGY LEVEL CLOSEST TO THE ATOMIC NUCLEUS.
FIRST LEVEL - 2 ELECTRONS COMPLETE
SECOND LEVEL - 8 ELECTRONS

ILLUSTRATIVE WAY OF INDICATING THAT ELECTRONS MOVE WITH THE ENERGY LEVEL.

SPECIFIC REGIONS OF SPACE WITHIN AN ENERGY LEVEL IN WHICH ELECTRONS OCCUR.

COMBINATION OF TWO OR MORE DIFFERENT ELEMENTS IN A SPECIFIC RATIO.

COMBINATION OF TWO OR MORE ATOMS.

SHOWS THE TYPES AND NUMBERS OF ATOMS IN A MOLECULE AND THEIR PARTICULAR ARRANGEMENT.

SHORTHAND METHOD TO DESCRIBE CHEMICAL COMPOSITION OF MOLECULE.

THESE ARE ATOMS THAT HAVE THEIR OUTER ELECTRON ORBITAL FILLED AND ARE THUS NON-REACTIVE OR CHEMICALLY INERT. (HE, NE)

<u>REACTANTS</u> ON THE LEFT SIDE OF THE EQUATION, <u>PRODUCTS</u> ON THE RIGHT SIDE. THIS IS A CONVENIENT MEANS OF SHOWING PRECISELY HOW MANY ATOMS OR MOLECULES PARTICIPATE IN A PARTICULAR CHEMICAL REACTION. THE NUMBERS INDICATE HOW MANY ATOMS OR MOLECULES OF EACH TYPE ARE INVOLVED.

ATTRACTIVE FORCE THAT HOLDS TWO ATOMS TOGETHER. VALENCE ELECTRONS DICTATE HOW MANY BONDS CAN FORM. (GENERAL IDEA.)

WHEN OUTER MOST SHELL OF AN ATOM CONTAINS LESS THAN LIGHT, IT WILL LOSE, GAIN OR SHARE ELECTRONS TO GET LIGHT ELECTRONS IN ITS OUTER SHELL.

COVALENT BOND

MOLECULE

SINGLE COVALENT BOND

DOUBLE BOND
(DOUBLE CONVALENT BONE)

ELECTRONEGATIVITY

POPLAR COVALENT BOND

NONPOLAR COVALENT BOND

IONIC BONDS

CATIONS

ANION

IONIC BOND

HYDROGEN BONDS

COMBINATION OF TWO OR MORE ATOMS JOINED BY COVALENT CHEMICAL BONDS.

SPECIFIC KIND OF CHEMICAL BOND. THE SHARING OF ELECTRONS BETWEEN ATOMS.

TWO PAIRS OF ELECTRONS ARE SHARED BY TWO ATOMS.

ONE PAIR OF ELECTRONS SHARED BY TWO ATOMS.

COVALENT BOND BETWEEN ATOMS WHICH HAVE DIFFERENT ELECTRONEGATIVITIES.

MEASURE OF AN ATOM'S ATTRACTIVE FORCE FOR THE ELECTRONS IN CHEMICAL BONDS.

ELECTRONS ARE PULLED FROM ONE ATOM TO THE OTHER. NUMBER OF PROTONS STAYS THE SAME - THUS GENERATE AN ION.

ELECTRONS SHARED EXACTLY BY THE ATOMS.

ATOMS WHICH 5, 6 OR 7 OUTER SHELL ELECTRONS TEND TO GAIN ELECTRONS FROM OTHER ATOMS. THEY ARE NEGATIVELY CHARGED IONS.

ATOMS WHICH LOST ITS VALENCE ELECTRONS - HAS A POSITIVE CHARGE - IS A POSITIVE ION.

BOND FORMED WHEN HYDROGEN ATOM IS ATTRACTED TO TWO OTHER ATOMS - USUALLY OXYGEN AND NITROGEN - AND ANOTHER ELECTRONEGATIVE ATOM. CAN BE ONE OR MORE MOLECULES INVOLVED. THESE ARE WEAK BONDS.

FORCE OF ELECTRICAL ATTRACTION BETWEEN TWO OPPOSITELY CHARGED IONS. FORM IS COMPOUND. THESE BONDS OCCUR BETWEEN ELECTRON DONORS AND ELECTRON ACCEPTORS.

OXIDATION

REDUCTION

INORGANIC COMPOUNDS

ORGANIC COMPOUNDS

POLAR

SOLVENT

ADHESIVE AND COHESIVE PROPERTIES

TEMPERATURE STABILIZATION

ACID

BASE

PH

SALT

CHEMICAL PROCESS IN WHICH A SUBSTANCE GAINS ELECTRONS.

CHEMICAL PROCESS WHERE A SUBSTANCE LOSES ELECTRONS.

LARGE, COMPLEX, CARBON-CONTAINING SUBSTANCES.

SMALL, SIMPLE SUBSTANCES

WATER IS AN EXCELLENT SOLVENT - IT IS CAPABLE OF DISSOLVING MANY POLAR SUBSTANCES.

WATER IS PARTIALLY POSITIVELY AND PARTIALLY NEGATIVELY CHARGED. WATER MOLECULE CAN HYDROGEN BOND WITH AS MANY AS FOUR OTHER WATER MOLECULES.

WATER REQUIRES A LOT OF HEAT TO CHANGE ITS TEMPERATURE AND THEREFORE HAS A HIGH SPECIFIC HEAT. THIS, TOO, IS THE RESULT OF THE HYDROGEN BONDS.

WATER MOLECULES STICK TO ONE ANOTHER BY HYDROGEN BONDS - <u>COHESIVE</u> WATER MOLECULES ALSO STICK TO OTHER CHARGED ATOMS OR MOLECULES - AND THE WATER <u>WETS</u> THESE SUBSTANCES. THIS IS THE ADHESIVE PROPERTY OF WATER.

PROTON ACCEPTOR. YIELD AN OH AND A CATION WHEN DISSOLVED IN WATER.

COMPOUND THAT IONIZES IN SOLUTION TO FORM A HYDROGEN ION (H+) AND AN ANION - THUS IT IS A PROTONE (H+) DONOR.

FORMED WHEN ACID AND BASE MIXED TOGETHER. COMPOUND IN WHICH THE HYDROGEN ATOM OF AN ACID IS REPLACED BY ANOTHER CATION.

WAY OF INDICATING THE DEGREE OF ACIDITY OR ALKALINITY. SCALE FROM 0 TO 14.

BUFFERS

SUBSTANCES WHICH CAN RESIST CHANGES IN PH WHEN ACIDS OR BASES ARE ADDED.

CHAPTER 3

THE CHEMISTRY OF LIFE: ORGANIC COMPOUNDS

BEFORE READING ACTIVITIES

I. Before you begin to read Chapter 3, review the highlights of Chapter 2. Explain briefly the following terms: chemical elements, atoms, compounds, bonds, oxidation and reduction, and inorganic compounds.

List the important groups of inorganic compounds.

II. Look over the outline below and answer the questions which follow it. Again, remember, you are NOT supposed to know all the answers before you read the chapter.

- I. Carbohydrates
  - A. Monosaccharides
  - B. Disaccharides
  - C. Polysaccharides
- II. Lipids
  - A. Neutral fats
  - B. Phospholipids
  - C. Steroids
- III. Proteins
  - A. Amino acid structure
  - B. Protein structure - levels of organization
- IV. Nucleic acids and nucleotides

Focus on classifying proteins

Questions

1. What are the basic groups of organic compounds?

2. How do you think the organic compounds discussed in this chapter relate to the inorganic compounds discussed in Chapter 2?

3. What are the subgroups of carbohydrates? Use your knowledge of word parts (mono-, di-, poly-, saccharide) to decide how these groups differ from each other.

4. Have you heard of these groups of organic compounds before? If so, in what context?

III. Read the summary on pages 58 and 59. How does the information stated in the summary fit in with information you already knew?

IV. Vocabulary

1. Review the vocabulary from Chapter 2 by giving yourself a test. If you missed some terms, separate these cards and continue to study them.

2. Sort the terms on the vocabulary cards into groups which you think will fit into the major and sub-sections of the chapter.

3. How many of these terms do you already know?

V. Flip through the pages of this chapter (45 through 58). How many figures are included in the chapter? What do these figures depict? How many parts are there in each figure?

DURING READING ACTIVITIES

I. Remember to read the chapter from one major section to the next.

II. Vocabulary

1. Circle the vocabulary terms in your text.

2. Use the vocabulary cards to make sure that you understand the definition.

3. Be sure to make vocabulary cards for any terms which are new to you and not on the printed cards.

III. Marking the text

1. Mark the most information by highlighting or underlining. Be sure to read the whole paragraph BEFORE you make any parts of the page.

The example below is from page 48 of the textbook.

A. Only key words necessary to understand the meaning of the portion marked are underlined. It is NOT NECESSARY to mark whole sentences.

B. This paragraph has MORE THAN ONE important idea. It is not enough to mark only the first sentence.

C. Since the examples of polysaccharides are important, they are listed in the margin.

Polysaccharides

The most abundant carbohydrates are polysaccharides, such as starches, glycogen, or celluloses. A polysaccharide is a single long chain, or a branched chain, consisting of repeating units of a simple sugar, usually glucose. The precise number of sugar units varies, but typically, thousands of units may be present in a single molecule of a polysaccharide. Because they are composed of different stereoisomers of glucose, or because the glucose units are arranged differently, these polysaccharides--starch, glycogen, and celluloses--have very different properties.

IV. Reading Figures

1. Note that each figure in this chapter is divided into two or three parts. For each figure you should understand what each part represents and how each part differs. For example, turn to Figure 3-3 on page 51.

   Part A illustrates the molecular structure of two compounds: glycerol (3-carbon alcohol) and a fatty acid.
   Part B illustrates the yield of the hydrolosis of a triacylglycerol.
   Part C depicts a common structure made up of neutral fats.

2. Be sure you look at the non-verbal portion of each figure and try to explain what it means before you read the verbal explanation.

3. Read the verbal explanation of the figure and compare it with your own explanation.

4. Consider how this figure helps to clarify information already presented in the text.

5. If you will need to be able to construct figures like these, start to practice now.

AFTER READING ACTIVITIES

I. Review your textbook marking. Remember, one reason for marking your testbook is to cut down the amount of reading you must do when you review for an examination.

II. Review the vocabulary

1. Go through all the terms you circled during reading and match them with the vocabulary cards. Repeat the definitions and sort the cards into those you know and those you need to learn.

2. Study the terms you need to learn in groups of 5-7 words for each day. Practice each new group several times each day. Review old words before you begin to learn new words. Include words learned in Chapter 2.

III. Answer the questions which follow.

1. Complex chemical compounds which contain carbon and make up the structural and functional components of cells and tissues are ________________ ________________.

2. Compounds, which are waxy or greasy in consistency and insoluble in water, belong to a diverse population of molecules called ________________.

3. Water loving molecules are said to be ________________, while water hating molecules are called ________________.

4. The building blocks of nucleic acids are ________________.

5. The building blocks of proteins are __________ ______________.

6. Proteins are composed of _____________ _____________ linked together by strong, covalent bonds called ___________________ _____________.

7. Since our bodies cannot make (synthesize) all twenty different amino acids, we must eat foods which contain the _____________ _____________ _____________.

8. The sequence of amino acids in a protein constitutes the _____________ _____________ of the protein.

9. _____________ is the storage polysaccharide in animal cells, while _____________ is the storage carbohydrate in plants.

10. All sugars and starches which contain carbon, hydrogen, and oxygen atoms are _____________.

11. Single long chains of repeating units of simple sugars, such as glucose, are _____________.

12. Lipid molecules which consist of carbon atoms arranged in rings are _____________. The molecules include cholesterol and the female and male sex hormones.

13. The biologically important molecule resulting from the linking together of amino acids by peptide bonds is a _____________.

14. In the space provided next to each compound below, place an L if the compound is a lipid, a C if the compound is a carbohydrate, or an N if the compound is a nucleic acid.

A. _______ DNA
B. _______ Hexose
C. _______ Cholesterol
D. _______ Phospholipid
E. _______ Glucose
F. _______ Pentose
G. _______ Starch
H. _______ RNA
I. _______ Steroid
J. _______ Polysaccharide

15A. Do all proteins necessarily have primary, secondary, and tertiary structure?

15B. Do all proteins have primary, secondary, tertiary, and quaternary structure?

16. What is special about the atomic structure of the carbon atom that enables it to form so many organic compounds?

17. Molecules have biological activity which is dependent on the _____________ _____________ structure of these molecules.

18. Identify the three different types of carbohydrates.

A. ________________
B. ________________
C. ________________

19. Distinguish between hydrophobic and hydrophilic molecules.

20. List the three important biological functions of lipids.

A. ___________________________
B. ___________________________
C. ___________________________

21. Which of the carbohydrates is the most abundant in the world?

22. Vegetable oil is an example of an ________________ ______________ ______________.

23. On page 54 examine the Focus on Classifying Proteins and answer the following questions.

A. What are the biological functions of proteins?

B. What are the two groupings of proteins when they are classified according to solubility?

Chapter 3 - Answers

1. Organic compounds
2. Lipids
3. Hydrophilic, hydrophobic
4. Nucleotides
5. Amino acids
6. Amino acids, peptide bonds
7. Essential amino acids
8. Primary structure
9. Glycogen, starch
10. Carbohydrates
11. Polysaccharides
12. Steroids
13. Polypeptide
14. Lipid, carbohydrate, nucleic acid
    A. N
    B. C
    C. L
    D. L
    E. C
    F. C
    G. C
    H. N
    I. L
    J. C

15A. Yes

15B. No. Only those proteins composed of two or more individual polypeptide subunits have quaternary structure.

16. Carbon has four electrons in its outer energy level, and therefore can participate in many chemical reactions.
17. Three dimensional
18. A. Monosacchorides
    B. Disacceharides
    C. Polysaccaharides
19. Hydrophobic are water-hating molecules, and hydrophilic molecules are water-loving.
20. A. Form cell membranes
    B. Biological fuel molecules
    C. Essential hormones
21. Cellulose because plants are so plentiful
22. Unsaturated fatty acid

23A. Enzymes, structural proteins, contractile proteins, hormones, transport proteins, and defense proteins

23B. Globular and fibrous proteins

ORGANIC COMPOUNDS

CARBOHYDRATES

MONOSACCHARIDE

HEXOSE

GLUCOSE

PENTOSE

DISACCHARIDE

POLYSACCHARIDES

STARCH

GLYCOGEN

CHITIN

LIPIDS

SUGARS AND STARCHES. CONTAIN CARBON, HYDROGEN AND OXYGEN ATOMS IN RATIO OF ONE TO TWO TO ONE.

COMPLEX CHEMICAL COMPOUNDS THAT CONTAIN CARBON. THESE COMPOUNDS ARE THE STRUCTURAL COMPONENTS OF CELLS AND TISSUES, PARTICIPATE IN METABOLIC REACTIONS AND ARE THE FUEL MOLECULES.

A SIX CARBON SUGAR WITH THE GENERAL FORMULA OF $C6H12O6$. GLUCOSE AND FRUCTOSE ARE COMMON HEXOSES.

SIMPLE SUGARS WHICH CONTAIN FROM THREE TO SIX CARBONS.

FIVE CARBON SUGARS SUCH AS RIBOSE AND DEOXYRIBOSE.

HEXOSE. OFTEN REFERRED TO AS BLOOD SUGAR AND IS UTILIZED IN THE CELL AS THE FUEL SUGAR.

SINGLE LONG CHAIN OR BRANCHED CHAIN WHICH CONSISTS OF REPEATING UNITS OF A SIMPLE SUGAR - FREQUENTLY GLUCOSE. INCLUDES STARCHES, GLYCOGEN AND CELLULOSE.

A SUGAR MOLECULE WHICH CAN BE DEGRADED INTO TWO MONOSACCHARIDE UNITS. INCLUDES MALTOSE, SUCROSE, AND LACTOSE.

POLYSACCHARIDE OF GLUCOSE THAT IS STORED IN ANIMAL TISSUE.

THE STORAGE CARBOHYDRATE IN PLANTS.

HETEROGENEOUS GROUP OF COMPOUNDS WHICH ARE GREASY OR OILY IN CONSISTENCY AND ARE INSOLUBLE IN WATER. ALSO CONTAIN CARBON, HYDROGEN AND OXYGEN.

A STRUCTURALLY STRONG POLYSACCHARIDE WHICH FORMS THE EXTERNAL SKELETONS OF INSECTS AND THE OTHER ARTHROPODS.

NEUTRAL FATS

GLYCEROL

FATTY ACID

PHOSPHOLIPID

HYDROPHOBIC

HYDROPHILIC

STERIOD

PROTEINS

AMINO ACIDS

PEPTIDE BOND

POLPEPTIDE

ESSENTIAL AMINO ACIDS

THREE CARBON ALCOHOL. (SEE FIGURE 3-8)

ABUNDANT LIPID IN LIVING THINGS. EFFICIENT FUEL MOLECULE. CONSISTS OF GLYCEROL JOINED TO TWO OR THREE MOLECULES OF FATTY ACID.

CONSISTS OF A GLYCEROL MOLECULE ATTACHED TO ONE OR TWO FATTY ACIDS. ALSO, THE GLYCEROL IS BONDED TO A PHOSPHOROUS WHICH IS A PART OF AN ORGANIC BASE. THESE MOLECULES ARE FOUND IN BIOLOGICAL MEMBRANES.

LONG, STRAIGHT CHAIN OF CARBON AND HYDROGEN ATOMS. (SEE FIGURE 3-8)

WATER LOVING MOLECULE.

WATER-HATING OR WATER AVOIDING MOLECULE.

BIOLOGICALLY IMPORTANT MOLECULES WHICH HAVE MANY STRUCTURAL AND ENZYMIC FUNCTIONS. THESE MOLECULES CONSIST OF SOMETIMES HUNDREDS OF AMINO ACIDS IN A SPECIFIC ORDER LINKED TOGETHER BY PEPTIDE BONDS.

MOLECULE CLASSIFIED AS A LIPID BUT DIFFERENT FROM OTHER LIPIDS. CONSISTS OF CARBON ATOMS ARRANGED IN INTERLOCKING RINGS: THREE RINGS CONTAIN SIX CARBON ATOMS AND THE FOURTH CONTAINS FIVE.

THE COVALENT BOND LINKING TWO AMINO ACIDS.

THE BUILDING BLOCK UNITS OF PROTEINS - THERE ARE TWENTY DIFFERENT AMINO ACIDS WHICH MAKE UP ALL PROTEINS.

THE AMINO ACIDS WHICH AN ANIMAL CANNOT SYNTHESIZE AND MUST TAKE IN VIA ITS DIET.

A POLYMER CHAIN OF AMINO ACIDS EACH LINKED TO THE NEXT BY A PEPTIDE BOND.

PRIMARY STRUCTURE OF PROTEINS

SECONDARY STRUCTURE OF PROTEINS

TERTIARY STRUCTURE OF PROTEINS

DENATURATION

QUATERNARY STRUCTURE OF PROTEINS

DEOXYRIBONUCLEIC ACID - DNA

RIBONUCLEIC ACID-RNA

NUCLEOTIDES

THE BENDING AND COILING OF A POLYPEPTIDE CHAIN WHICH GIVES THE MOLECULE SOME THREE DIMENSIONAL SHAPE, BUT THIS IS NOT THE FINAL CONFORMATION OF THE MOLECULE.

THE SEQUENCE OF AMINO ACIDS IN A POLYPEPTIDE CHAIN.

THE COMPLETE UNFOLDING OF THE CONFORMED PROTEIN MOLECULE WHICH RESULTS IN A LOSS OF BIOLOGICAL ACTIVITY.

THE THREE DIMENSIONAL CONFORMATION OF A PROTEIN MOLECULE RESULTING FROM FURTHER FOLDING OF THE HELICAL AND PLEATED POLYPEPTIDE. STABILIZED BY WEAK CHEMICAL BONDS.

NUCLEIC ACID WHICH MAKES UP THE GENES, THE HEREDITARY MATERIAL OF THE CELL. THESE DNA GENES CONTAIN THE ENCODED INFORMATION FOR MAKING ALL THE PROTEINS NEEDED BY THE ORGANISM.

SOME FUNCTIONAL PROTEINS ARE COMPOSED OF TWO OR MORE INDIVIDUAL POLYPEPTIDE CHAINS (EACH OF WHICH HAS PRIMARY, SECONDARY AND TERTIARY STRUCTURE WHICH COMBINE TO FORM THE BIOLOGICALLY ACTIVE MOLECULE).

THE BUILDING BLOCKS OF NUCLEIC ACIDS WHICH ARE THEMSELVES COMPOSED OF A FIVE CARBON SUGAR, A NITROGENOUS BASE, AND A PHOSPHATE GROUP. (SEE FIGURE 3-13)

NUCLEIC ACID WHICH MAKES UP SOME OF THE STRUCTURES INVOLVED IN PROTEIN SYNTHESIS.

CHAPTER 4

CELL STRUCTURE AND FUNCTION

BEFORE READING ACTIVITIES

I. Before you begin to read Chapter 4, review the highlights of Chapters 2 and 3.

1. Explain the difference between organic and inorganic compounds.

2. What are atoms?

3. What are important groups of inorganic compounds?

4. What are the important groups of organic compounds?

II. Look over the outline below and answer the questions which follow it.

CELL STRUCTURE AND FUNCTION

I. General characteristics of cells
II. How cells are studied
III. Prokaryotic and eukaryotic cells
IV. Inside the cell
    a. Endoplasmic reticulum and ribosomes; internal transport and synthesis
    B. The Golgi complex: a packaging plant
    C. Lysosomes: breaking down harmful materials
    D. Microbodies
    E. Mitochondria: power plants of the cell
    F. Plastids: energy traps and storage sacs
    G. Microtubules and microfilaments: cell shape and movement
    H. The microtrabecular lattice: intracellular framework
    I. Centrioles
    J. Cilia and flagella
    K. Vesicles and vacuoles
    L. The cell nucleus
        1. The nuclear envelope
        2. The nucleolus
        3. Chromatin and chromosomes
    Focus on viewing the cell

1. How many types of cells are there? What are they? How do you think they differ?

2. Are cells simple or complex structures? How can you tell this? What are the internal cell structures?

III. Read the introductory section on page 62 and the Chapter Summary on page 83. Take the Post-Test on pages 83-84.

IV. Vocabulary

1. Review the vocabulary from Chapters 2 and 3. Be sure to continue studying terms you missed.

2. Read through the vocabulary terms for this chapter. Note the following:

A. There are two words which you think you already know, smooth ER, rough ER. Be sure to check their meaning.

B. The prefix "micro-" is used several times. What does it mean?

C. How does the prefix "mito-" differ from "micro-"?

D. What is the relationship of the terms which have "nucleus" or "nucleo" in them?

3. Sort the vocabulary cards into groups which fit into the major and subsections of the chapter.

V. Flip through the pages of this chapter (61-82).

1. Note that there is one table and a large number of figures.

2. Compare your vocabulary terms with the cell structures listed in Table 4-1.

3. Most of the figures depict cell structures. Which of these do you think you will need to be able to reproduce?

DURING READING ACTIVITIES

I. Continue to read the chapter from one major section to the next.

II. Vocabulary

1. Be sure you circle the vocabulary terms in your text.

2. Check to make sure that you found each term in the text that is on your vocabulary cards.

III. Marking the text

1. Continue to read the sections within the larger sections in whole paragraphs. Mark each paragraph after you read it completely.

The example below is from page 63 of the textbook.

(a) Again, only keywords are marked. You do NOT NEED to mark whole sentences and you do NOT want to mark the whole paragraph.

(b) These paragraphs have more than one important piece of information in each of them. Be sure you mark all important information.

(c) You should have circled the vocabulary terms: organelles, light microscope, electron microscope. Note that these terms are in darker type.

(d) Information which should be reviewed and remembered is written in the margins.

HOW CELLS ARE STUDIED

Cells are so small that one might well wonder how we know so much about what goes on inside them. The biologist's most important tool for studying the internal structure of cells has been the microscope. Anton van Leeuwenhoek (1632-1723) is credited with developing some of the earliest microscopes and with leaving written records of the structures he studied. In 1665 Robert Hooke (1635-1703) examined a slice of cork with the aid of a crude, homemade microscope. Because the tiny compartments he saw reminded him of the little rooms, or cells, of a monastery, he called them cells. What Hooke saw were the cell walls of dead cork cells [Figure 4-3]. In later observations he described cell contents, but it was not until two centuries later that scientists realized that the important part of the cell is its contents and not its outer walls or membranes. During the 1800s scientists studied various cells and observed a variety of intracellular structures, which they referred to as [organelles] (little organs) because they were thought to perform special jobs within the cell, just as our organs perform specific jobs within our bodies.

The orginary compound light microscope, the kind used by students in most college laboratories was responsible for the discovery of most cell structures. [The light microscope] gradually improved since Hooke's time, uses visible light as the source of illumination [See Focus on Viewing the Cell]. During the last three decades, the development of the electron microscope has enabled researchers to study the fine detail (ultrastructure) of cells. The [electron microscope] floods the specimen being studied with a beam of electrons rather than with light waves.

IV. Reading Figures

1. Again, note that the figures have more than one part. Be sure you understand what each part represents, how the parts are alike and how they differ. For example, turn to the Figure 4-8 on page 70.

   1. Both parts are structures of animal cells. Part A is a reproduction of an electron micrograph and part B is a drawing.

   2. Part A illustrates zymogen granules and discusses enzymes within and released from these granules.

   3. Part B points out desmosomes.

4. Cover the labels in the outer portion of the figures and test yourself on what name goes with each line.

5. Consider how this figure helps to clarify information already presented in the text.

AFTER READING ACTIVITIES

I. Review your textbook marking. Be sure that you have marked the important parts of each paragraph. Check to make sure that you understand what you have marked. Reread your margin notes. If you have forgotten what they mean, rewrite them so you can use them later for review.

II. Review the vocabulary.

1. Sort the cards into groups that fit together in some way. For example, all the words that begin with mirco- or all the membranous intracellular structures.

2. Take a practice test to determine which words you already know and which ones you need to study. Practice the words you need to learn in small groups of 5-7. Be sure that you practice each group several times.

III. Answer the questions which follow.

1. The large variety of intracellular structures which perform various important functions are cellular ____________.

2. Organisms which do not have a well-defined nucleus are called ____________.

3. The ____________ ____________ of a cell is a lipid bilayer with interspersed proteins. This structure regulates the movement of materials into and out of the cell.

4. The ____________ ____________ enables researchers to see the greatest details of cell structure.

5. The ____________ ____________ ____________ is the membranous system in the cell which has ribosomes attached to it.

6. ____________ are membranous sacs which contain oxidative enzymes.

7. ____________ are the power plants of the cell. The inner membrane of this structure is folded into ____________.

8. The ____________, ____________, and microtrabecular lattice work are the major elements of the cytoarchitecture.

9. The ____________ is a large, spherical structure which contains the chromatin material (DNA plus proteins). This is the information repository of the cell.

10. The cells contain membranous sacs filled with digestive enzymes. These are called ____________. They are formed in the cell by the ____________ ____________.

11. In chloroplasts the disc-like structures which are stacked into grana are called the ____________.

12. The flattened membranous sacs which are arranged in stacks of about eight, and which function to package proteins are called the ______________ ______________. This structure is named for the Italian scientist ______________.

13. List 5 membranous intracellular structures

    1. ____________________
    2. ____________________
    3. ____________________
    4. ____________________
    5. ____________________

14. Prokaryotic cells are distinguished from ______________ cells by the presence of a nucleus and internal membranous structures in these latter type cells.

15. How big is an average cell?

16. Cells vary dramatically in __________, ________, and function.

17. Plant and animal cells differ from one another in many ways. List the four major differences between these two major groups of cells.

    A.

    B.

    C.

    D.

18. Examine Figure 4-8. Be certain you can identify the major structures of the cell. Prepare a labeled diagram showing the major cell structures.

19. Examine Figure 4-9. Diagram a typical plant cell and label the major structures.

20. Examine Figure 4-10. Based on the diagram in Figure 4-10a, write a short and accurate description of the rough endoplasmic reticulum.

21. Examine Figure 4-11. In your own words, describe what is occurring in diagrams a through d.

22. Consider the fact that cellular energy production occurs on the inner mitochodrial membrane or cristae. What is the advantage to the cell of having these highly folded inner mitochondrial membranes?

23. Do mitochondria have their own DNA, and do they synthesize their own proteins?

24. How do microtubules change their length?

25. Distinguish between cilia and flagella.

26. Examine Figure 4-17. Rank the following subcellular structures in order of decreasing size: (a) mitochondria, (b) microtrabecular strand, (c) endoplasmic reticulum tubule, (d) microtubule.

27. Examine Figure 4-22. What is the probable function of nuclear pores?

28. If when you were looking at living cells under the light microscope you could see chromosomes, what can you conclude the cell is doing?

Chapter 4 - Answers

1. Organelles
2. Prokaryotes
3. Cell or plasma membrane
4. Electron microscope
5. Rough endoplasmic reticulum
6. Microbodies
7. Mitochondria
8. Microtubules, microfilaments
9. Nucleus
10. Lysosomes, Golgi complex
11. Thylakoids
12. Golgi complex, Golgi
13. 1. Mitochondria
    2. Chloroplasts
    3. Microbodies
    4. Lysosomes
    5. Golgi complex
    6. Nucleus
    7. Rough endoplasmic reticulum
    8. Smooth endoplasmic reticulum
14. Eukaryotic
15. 10 micrometers (1/2500 inch)
16. Size, shape
17. A. Both types of cells have cell membranes, but plant cells have a cellulose cell wall.
    B. Plant cells have membrane bound plastids which produce and store food material.
    C. Most plant cells have vacuoles which store and transport nutrients and waste products.
    D. Centrioles and lysosomes are absent in plant cells.
22. Increased surface area allows for greater production of cellular energy.
23. Yes, mitochondria have their own DNA; and yes there is a small number of proteins synthesized in the mitochondria.
24. Addition of tubulin subunits at one end of the microtubule, and disassembly of tubulin subunits at the other end.
25. Both are movable structures projecting from a free surface. If a cell has a few and they are very long, they are flagella. If they are short and plentiful, they are cilia.
26. a, c, d, b
27. Selective movement of materials into and out of the nucleus.
28. Cell is dividing. Would not see chromosomes in nondividing cells.

CELL

CELL THEORY

ORGANELLES

LIGHT MICROSCOPE

ELECTRON MICROSCOPE

RESOLVING POWER

PROKARYOTIC CELLS

EUKARYOTIC CELLS

CELL MEMBRANE

ENDOPLASMIC RETICULUM (ER)

SMOOTH ER

ROUGH ER

SCHLEIDEN AND SCHWANN PROPOSED THAT ALL LIVING THINGS ARE MADE UP OF CELLS; THE CELL IS THE BASIC UNIT OF THE LIVING ORGANISMS. VIRCHOW LATER ADDED THAT ALL CELLS DERIVED FROM OTHER CELLS BY DIVISION.

THE SMALLEST UNIT OF LIVING MATERIAL CAPABLE OF CARRYING ON ALL THE ACTIVITIES NECESSARY FOR LIFE; AND CELLS ARE THE BUILDING BLOCKS OF THE ORGANISM.

ORGINARY MICROSCOPE USING LIGHT TO ILLUMINATE STRUCTURES TO BE STUDIED. MAGNIFIES STRUCTURES ABOUT 1,000 TIMES. CAN RESOLVE OBJECTS 500 TIMES BETTER THAN THE HUMAN EYE.

THE VARIETY OF INTRACELLULAR STRUCTURES WHICH PERFORM SPECIAL JOBS IN THE CELLS.

THE ABILITY TO REVEAL FINE DETAIL, EXPRESSED AS THE MINIMUM DISTANCE BETWEEN TWO POINTS THAT CAN BE DISTINGUISHED AS SUCH.

USES ELECTRONS AND SPECIAL OPTICS TO MAGNIFY OBJECTS BEING STUDIED 250,000 TIMES. THE RESOLVING POWER IS 10,000 TIMES THAT OF THE HUMAN EYE.

CELLS WHICH HAVE A TRUE NUCLEUS (MEMBRANE BOUND).

BACTERIA AND CYANOBACTERIA. CELLS ARE SIMPLER AND SMALLER THAN EUKARYOTIC CELLS. THESE CELLS LACK A NUCLEAR MEMBRANE, AND THEIR DNA IS USUALLY CONCENTRATED IN NUCLEOIDS. THESE CELLS ALSO LACK OTHER MEMBRANE-BOUND ORGANELLES.

NETWORK OF INTERNAL MEMBRANES THROUGHOUT THE CYTOPLASM; FORMS EXTENSIVE NETWORK OF TUBULES AND VESICLES. SITE OF MANY CHEMICAL REACTIONS.

LIPID BILAYER WITH DIVERSE PROTEINS INTERSPERSED. REGULATES MOVEMENT OF MATERIALS INTO AND OUT OF THE CELL. MAINTAINS CELL SHAPE AND FUNCTIONS IN CELL.

ENDOPLOOMIC RETICULUM WHICH HAS ASSOCIATED RIBOSOMES. INVOLVED IN THE SYNTHESIS OF PROTEINS DESTINED FOR EXPORT.

MEMBRANEOUS NETWORK WHICH <u>LACKS RIBOSOMES</u>: IN LIVER DETOXIFIES DRUGS AND OTHER SUBSTANCES, AND IN OTHER CELLS SUCH AS ADRENAL GLAND PRODUCES STERIODS. INVOLVED IN LIPID METABOLISM.

RIBOSOMES

GOLGI COMPLEX

MICROBODIES

VESICLES AND VACUOLES

MITOCHONDRIA

PLASTIDS (CHLOROPLAST)

MICROTUBULES

MICROFILAMENTS

MICROTRABECULAR

CENTRIOLES

CILIA

FLAGELLA

FLATTENED MEMBRANOUS SACS ARRANGED IN STACKS. PACKAGES PROTEINS FOR SECRETION. FORMS LYSOZYMES.

STRUCTURES COMPOSED OF PROTEINS AND RNA. SOME ARE FREE IN CYTOPLASM AND OTHERS ARE BOUND IN ENDOPLASMIC RETICULUM. SITE OF PROTEIN SYNTHESIS.

MEMBRANOUS SACS; VESICLE IS SMALL VACUOLE. FUNCTIONS TO TRANSPORT AND STORE INGESTED MATERIALS, CELLULOR SECRETIONS, OR WASTES.

MEMBRANOUS SACS CONTAINING OXIDATIVE ENZYMES.

MEMBRANOUS STRUCTURES; CHLOROPLAST CONTAINS CHLOROPHYLL AND IS INVOLVED IN PHOTOSYNTHESIS. HIGHER PLANT CELLS HAVE 20-40 CHLOROPLASTS PER CELL.

STRUCTURE CONSISTING OF AND SMOOTH OUTER MEMBRANE AND A HIGHLY FOLDED INNER MEMBRANE. THESE FOLDS ARE THE CRISTAE. SITE OF SOME REACTIONS OF CELLULAR RESPIRATION AND PRODUCTION OF CELLULAR ENERGY. AS MANY AS 1,000 PER CELL.

PART OF CELL'S ARCHITECTURE (CYTOARCHITECTURE). CONSIST OF SOLID ROD-LIKE STRUCTURES COMPOSED OF THE PROTEIN ACTIN. STRUCTURAL SUPPORT AND CELL MOVEMENT.

PART OF THE CELL'S ARCHITECTURE (CYTOARCHITECTURE), HOLLOW RODS IN CYTOPLASM. COMPOSED OF PROTEIN TUBLIN. HAVE ROLE IN CELL MOVEMENT. LOCATED JUST BENEATH THE CELL MEMBRANE.

PAIR OF HOLLOW CYLINDERS LOCATED IN CENTROSOME REGION. EACH CENTRIOLE CONSISTS OF NINE TRIPLE MICROTUBULES. SERVES AT SITE FOR ORGANIZATION OF SPINDLE APPARATUS FOR CELL DIVISION.

ELEMENT IN CYTOARCHITECTURE. NETWORK OF SLENDER PROTEIN THREADS FORMING CELL FRAMEWORK. WORKS IN CONJUNCTION WITH MICROTUBULES AND MICROFILAMENTS TO CHANGE CELL SHAPE AND RESPONSES.

HOLLOW TUBES CONSISTING OF SMALLER MICROTUBULES; EXTEND OUTSIDE OF CELL CELLULAR LOCOMOTION. IN HUMANS FOUND <u>ONLY</u> IN SPERM CELLS.

HOLLOW TUBES CONSISTING OF MICROTUBULES. FUNCTION IN CELL MOVEMENT OR MOVEMENT OF MATERIAL OUTSIDE CELL.

NUCLEUS

NUCLEOLUS

CHROMOSOMES

CYTOPLASM

NUCLEOPLASM

LYSOSOMES

THYLAKOIDS

NUCLEAR ENVELOPE

ROUND STRUCTURE WITHIN THE NUCLEUS WHICH CONSISTS OF RNA AND PROTEINS. THIS IS THE SITE OF SYNTHESIS AND ASSEMBLY OF RIBOSOMES.

LARGE SPHERICAL STRUCTURE SURROUNDED BY NUCLEAR MEMBRANE, CONTAINS NUCLEOLUS AND THE DNA PLUS PROTEIN MATERIAL. INFORMATION REPOSITORY OF CELL.

JELLY-LIKE MATERIAL OF THE CELL EXCLUDING THE NUCLEUS. THE SUBCELLULAR ORGANELLES ARE SUSPENDED IN THE CYTOPLASM.

LONG THREAD-LIKE STRUCTURE COMPOSED OF DNA AND PROTEIN. THESE STRUCTURES ARE PRESENT WHEN THE CELL IS UNDERGOING MITOSIS AND MEIOSIS. THE DNA OF THE CHROMOSOMES IS THE GENETIC INFORMATION OF THE CELL.

MEMBRANOUS SACS CONTAINING DIGESTIVE ENZYMES. THESE ENZYMES, WHEN RELEASED, BREAK DOWN UNWANTED MATERIAL IN THE CELL. PLAY ROLE IN CELL DEATH. BREAKS DOWN BACTERIA AND OTHER DEBRIS.

JELLY-LIKE MATERIAL OF THE NUCLEUS

DOUBLE MEMBRANE STRUCTURE WHICH SURROUNDS THE NUCLEUS.

COMPONENTS OF CHLOROPLASTS WHICH ARE DISC-LIKE STRUCTURES, AND WHICH CONTAIN CHLOROPHYLL. THE LIGHT REACTIONS OF PHOTOSYNTHESIS TAKE PLACE IN THESE STRUCTURES. THE STACKS OF THYLAKOIDS ARE THE GRANA.

CHAPTER 5

THE CELL MEMBRANE: INTERACTING WITH THE ENVIRONMENT

BEFORE READING ACTIVITIES

I. Before you begin to read Chapter 5, review the major ideas in Chapter 4.

1. What are the general characteristics of cells? How does information presented in Chapter 5 fit into the information presented in Chapter 4?

2. What are the structures inside the cell?

3. What is the major function of the cell membrane? (Hint: Look at the title of Chapter 5.)

II. Look over the outline below and answer the questions which follow it.

THE CELL MEMBRANE: INTERACTING WITH THE ENVIRONMENT

I. Why do cells need cell membranes?
II. The structure of the cell membrane
  A. The lipid bilayer
  B. Membrane proteins
  C. Microvilli
III. Cell walls
IV. Cell junctions
V. How do materials pass through membranes?
  A. Diffusion
  B. Osmosis
  C. Active transport
  D. Endocytosis and exocytosis
Focus on splitting the lipid bilayer

1. What are the five major ideas that you need to know about cell membranes? (Use the Roman numerals in the outline to help you.)

2. What are three structures of the cell membrane?

3. List four ways materials pass through membranes. How do you think each is different?

4. Where did the outline above come from? (Check page 85 of your textbook.)

III. Read the short introductory paragraph on page 86 and the Chapter Summary on page 100. Take the Post-Test on pages 100-101.

IV. Vocabulary

1. Review the vocabulary from Chapter 4. Remember that these terms describe internal cell structures.

2. Read through the vocabulary cards for this chapter. Remember that these terms relate to the outer cell structure.

   A. The word part "micro-" is used again.

   B. Several of the vocabulary terms include some form of the word "permeable." How do these terms relate to each other.

   C. There are three terms which include the word part "_cytosis." How are these words alike and how do they differ?

   D. There are three terms which contain the word part "-tonic." How are these words alike and how do they differ?

V. Flip through the pages of Chapter 5 (86-99).

1. How do the BOLD FACE headings relate to the outline above? How do the lower case dark headings relate to the outline (The Lipid Bilayer). Where did the outline come from? Check page 85 of your textbook.

2. Note that there are two tables and 16 figures. The figures are an important part of this chapter. Be sure you use them as you read.

DURING READING ACTIVITIES

I. Read the chapter from one BOLDFACE heading to the next BOLDFACE heading. As you complete a section, stop and sum up what the whole section was about. For example, what is the answer to the question "Why do cells need cell membranes?" You should be able to give four reasons.

II. Vocabulary

1. Circle the vocabulary terms in your text; most of the terms new to you are in darker type, but some, such as asymmetrical, may not be. Be sure you circle any term you do not know and make a vocabulary card for it.

2. Underline the definitions given for terms you circle whenever they appear in the text.

III. Marking the text

1. Be sure that you read a whole paragraph or a group of paragraphs before you make any marks on the page. Be sure you use the margins to make notes.

2. The example below is from page 88 of your textbook.

   1. Note that the first sentence in the paragraph distinguishes between the lipid bilayer and the membrane protein.

   2. Be sure you know what "dynamic" means in this context.

3. Your margins may not look neat, but you will have the critical information available quickly when you begin to review this chapter.

MEMBRANE PROTEINS

1. Most proteins associated with inner surface (in cytoplasm)
2. Proteins that protrade glycoproteins away from cytoplasm
3. Glycoproteins function in communication

The lipid bilayer is a fluid matrix in which the proteins may move about. The lipids serve as a barrier to polar molecules and ions, whereas the membrane proteins carry out specific functions of the membrane, such as chemical transport, energy transfer and transmission of messages. Membranes may differ in the numbers and kinds of proteins present; membranes with different functions contain different proteins. Furthermore, cell membranes are [dynamic] structures, and their chemical composition and molecular arrangement may change with varying conditions.

Most of the membrane proteins are associated with the inner, cytoplasmic surface of the membrane [See Focus on Splitting the Lipid Bilayer]. The proteins that protrude from the outer surface (away from the cytoplasm) are mainly [glycoproteins], that is, proteins to which sugar residues are attached. Glycoproteins on the cell membrane appear to serve as the cell's communication system with its outer environmant, both with other cells and with messenger molecules, such as hormones.

difference between lipid bilayer and membrane proteins.

IV. Reading Figures

1. Compare the table on page 93, "Mechanisms for Moving Materials Through Cell Membranes" with the table on page 68, "Eukaryotic Cell Structures and Their Functions." Both list the terms in the column on the left. Most of these terms are part of your vocabulary for each chapter. The columns to the right give some explanations of each term. These columns also identify which things "go together" (for example the processes that have random molecular motion are alike in an important way.) The columns also show how things are different.

2. Look at Figure 5-10 on page 961 Before you read the explanation, try to decide what the illustrations depict. How is (a) different from (b)? from (c)? What is the difference between the top portion of the figure and the bottom portion? Is your explanation of this figure similar to that of the authors?

AFTER READING ACTIVITIES

I. Again, check your textbook marking. How mjch of what you underlined or wrote in the margins did you remember without having to read all of it? Did you have to correct your first markings less in this chapter than you did on Chapter 2?

II. Review the vocabulary.

1. Sort the cards into groups that fit together. Did you put all the terms with "cytosis" together? Did you make a card for "endocytosis"? What is alike about these processes? What is different?

2. Again, as always, use a practice test to determine which words you need to study.

III. Answer the questions which follow.

Questions

1. List the four major functions of the cell membrane.

2. Is the cell membrane seen with the light microscope? Explain.

3. The lipid components of the cell membrane all have an important common feature. These molecules have one highly polar _______________ portion and one highly _________________ hydrophobic portion.

4. The formation of the lipid bilayer is a ______________ process driven by the hydrophobic interactions of the hydrocarbon chains. No ______________ bonds are involved in formation of the lipid bilayer.

5. __________ ___________ are self-sealing and self-assembling.

6. Examine text Figure 5-2. Describe the orientation of the lipid molecules.

7. Examine text Figure 5-2. Describe the positioning of the membrane proteins.

8. Do the proteins move about in the lipid bilayer or are they restricted to a particular region?

9. What is a possible function for the microvilli in the intestine?

10. Which of the following cells have thick cell walls?

    A. Animal
    B. Fungi
    C. Algae
    D. Plants
    E. Bacteria

11. What substance is the main component of the cell wall?

12. What compounds are found in wood?

13. List the four different kinds of cell junctions.

A. ____________________
B. ____________________
C. ____________________
D. ____________________

14. Distinguish between gap junctions and plasmodesmata.

15. Distinguish between diffusion and osmosis.

16. Which processes of membrane transport require no cellular energy?

17. Which processes of membrane transport require the expenditure of cellular energy?

18. For each of the different mechanisms for moving materials across a cell membrane which are listed below, provide a biological example of that process.

Process	Example
A. Diffusion	________________________
B. Active Transport	________________________
C. Pinocytosis	________________________
D. Exocytosis	________________________
E. Osmosis	________________________

19. What factors affect the rate of diffusion of a substance?

20. Heat energy causes molecules to move more ____________.

21. In osmosis cell membranes allow the ________________ movement of water molecules across them.

22. Hydrophilic molecules cross the hydrophobic cell membrane through a channel formed by four or more transport ____________ ________________.

23. Larger quantities of materials are moved in and out of the cell by ________________ and ________________.

24. Examine text Figure 5-15. Once material is taken into a cell by phagocytosis, how is it utilized by the cell?

Chapter 5 - Answers

1. A. The cell membrane regulates the flow of materials into and out of the cell.
   B. The cell membrane receives information that permits the cell to sense changes in its environment and to respond to them.
   C. The cell membrane maintains structural and chemical relationships with neighboring cells.
   D. The cell membrane protects the cell; may function in secretion of substance; may be involved in cell movement; and in some cells it is involved in transmitting impulses.
2. No, the cell membrane is very thin 6-10 nanometers, and an electron microscope is required to see these membranes.
3. Hydrophilic, nonpolar
4. Spontaneous, covalent
5. Lipid bilayers
6. The polar heads are oriented towards the inside and outside surfaces of the membrane.
7. The membrane proteins are on both the inside and outside surfaces of the membrane, and some proteins span the entire membrane.
8. Proteins move about
9. Increase the surface area for absorption of materials
10. b, c, d, e, all have cell walls
11. Cellulose
12. Cellulose reinforced with lignin
13. A. Plasmodesmata
    B. Gap junctions
    C. Desmosomes
    D. Tight junctions
14. Gap junctions are the cell junctions in animal cells which allow the passage of materials between two adjacent cells. Plasmodesmata have the same function in plant cells.
15. Both processes refer to the movement of materials from region of high concentration to one of low concentration. However, osmosis refers only to the movement of water molecules.
16. Diffusion, facilitated diffusion, and osmosis
17. Active transport, phagocytosis, pinocytosis, and exocytosis
18. A. Movement of oxygen in tissue fluid
    B. Pumping sodium out of cells
    C. Cells take in needed solute and solution
    D. Secretion of mucus
    E. Water enters red blood cell
19. Size and shape of the molecule, the temperature and whether the medium is solid liquid or gas
20. Rapidly
21. Free, unrestricted
22. Protein molecules
23. Exocytosis, endocytosis
24. Lysosomes which contain breakdown enzymes, fuse with vesicle and digest the material

PERMEABLE

SELECTIVE PERMEABILITY

FLUID MOSAIC MODEL (CELL MEMBRANE)

GLYCOPROTEINS

MICROVILLI

PLASMODESMATA

GAP JUNCTIONS

DESMOSOMES

TIGHT JUNCTIONS

PERMEABLE MEMBRANE

IMPERMEABLE MEMBRANE

SELECTIVELY PERMEABLE MEMBRANE

CELL MEMBRANE PERMITS ONLY CERTAIN SUBSTANCES TO ENTER OR LEAVE THE CELL. SOME SUBSTANCES ARE NOT ALLOWED TO CROSS THE CELL MEMBRANE, WHILE OTHER SUBSTANCES ARE FACILITATED IN CROSSING.

SUBSTANCES PASS FREELY ACROSS THE CELL MEMBRANE. CELLS ARE NOT TOTALLY PERMEABLE TO ALL SUBSTANCES.

PROTEINS TO WHICH SUGAR MOLECULES ARE COVALENTLY LINKED.

THE MEMBRANE CONSISTS OF A LIPID BILAYER IN WHICH ARE EMBEDDED A VARIETY OF GLOBULAR PROTEINS. A LIPID BILAYER IS A DOUBLE LAYER OF LIPIDS.

CYTOPLASMIC EXTENSIONS JOINING TWO PLANT CELLS WHICH PROVIDE A PATHWAY FOR THE PASSAGE OF WATER, IONS, NUTRIENTS, AND OTHER MATERIALS.

TINY EVAGINATIONS OF THE CELL MEMBRANE. FUNCTION TO VASTLY INCREASE THE SURFACE AREA.

BUTTON-LIKE PLAQUE STRUCTURES WHICH PROVIDE THE ADHESION BETWEEN TWO CELLS WHICH ARE IN CLOSE PROXIMITY TO ONE ANOTHER.

CYTOPLASMIC EXTENSIONS BETWEEN ADJACENT ANIMAL CELLS WHICH ALLOW THE MOVEMENT OF WATER, IONS, NUTRIENTS AND OTHER MATERIALS.

MOVEMENT OF A GIVEN SUBSTANCE THROUGH A CELL MEMBRANE.

INTERCELLULAR JUNCTIONS WHICH ARE VERY TIGHT (NO INTERCELLULAR SPACE). THE CELL MEMBRANES ARE FUSED, AND PROVIDE A SHARP PHYSICAL SEPARATION BETWEEN BODY COMPARTMENTS.

A MEMBRANE WHICH ALLOWS THE PASSAGE OF SOME SUBSTANCES. THIS IS A PROPERTY OF CELLULAR MEMBRANES.

NO MOVEMENT OF GIVEN SUBSTANCE THROUGH A CELL MEMBRANE.

DIFFUSION

FACILITATED DIFFUSION

OSMOSIS

ACTIVE TRANSPORT

PHOGOCYTOSIS

PINOCYTOSIS

EXOCYTOSIS

DOWN A CONCENTRATION GRADIENT

ISOTOMIC

HYPERTONIC

HYPOTONIC

PROCESS WHEREBY CERTAIN SUBSTANCES ARE MOVED ACROSS A MEMBRANE BY MEMBRANE PROTEINS FROM A REGION OF HIGH CONCENTRATION TO A REGION OF LOW CONCENTRATION. NO CELLULAR ENERGY REQUIRED.

NET MOVEMENT OF MOLECULES FROM AREA OF HIGH CONCENTRATION TO AREA OF LOW CONCENTRATION. DEPENDS ON RANDOM MOLECULAR MOVEMENTS.

PROCESS WHEREBY MOLECULES OR IONS ARE TRANSPORTED ACROSS THE CELL MEMBRANE BY CELL MEMBRANE PROTEINS. CELLULAR ENERGY IS REQUIRED. MAY MOVE SUBSTANCES FROM A REGION OF LOW CONCENTRATION TO REGION OF HIGH CONCENTRATION.

DEFFUSION OF WATER MOLECULES FROM REGION OF HIGH CONCENTRATION TO REGION OF LOW CONCENTRATION. NO CELLULAR ENERGY REQUIRED.

MEANS "CELL DRINKING." PROCESS IN WHICH CELL MEMBRANE TAKES IN FLUID DROPLETS BY FORMING VESICLES AROUND THEM.

MEANS "CELL EATING." PROCESS WHEREBY PORTION OF CELL MEMBRANE SURROUNDS A PARTICLE AND BRINGS IT INTO THE CELL.

MOVEMENT FROM A HIGH TO A LOW CONCENTRATION.

PROCESS IN WHICH MATERIALS ARE EJECTED ACROSS THE CELL MEMBRANE; INVOLVES THE FUSION OF SECRETORY VESICLE MEMBRANES WITH THE CELL MEMBRANE.

WHEN A SOLUTION HAS A GREATER SOLUTE CONCENTRATION THAN THE CELL.

WHEN CELLS ARE PLACED IN A SOLUTION IN WHICH THE CONCENTRATION OF A SUBSTANCE (SOLUTE) IS THE SAME AS FOUND INSIDE THE CELL.

WHEN A SOLUTION HAS A LESSER CONCENTRATION THAN THE CELL.

CHAPTER 6

THE ENERGY OF LIFE

BEFORE READING ACTIVITIES

I. Before you begin to read Chapter 6, note that this is the first chapter in a new section. Review the important concepts in Section I.

1. How is life organized? You should think of atoms and molecules, organic compounds, cell structure and function, and the cell membrane.

2. How do atoms and molecules differ from organic compounds? from cell structure? from cell membranes?

3. What do you think is the relationship between Section I, "The Organization of Life" and Section II, "Life and the Flow of Energy"?

4. Look at the chapter headings for Part II on page ix of the Contents Overview. What are the most important concepts you will need to know about the life and flow of energy? Remember, each chapter will explain one of these concepts.

II. Getting an overview

1. Turn to page 105 of your textbook and look over the outline detailing the important concepts about the energy of life. Answer the following questions using the outline.

    1. What are the two types of energy?

    2. How do you think energy is measured?

    3. What do you think "equilibrium" means in this outline?

    4. Do you think enzymes are important? Your answer should be "yes" because a lot of the chapter is spent explaining about various aspects of enzymes.

2. Read the Learning Objectives listed below and answer the questions which follow it.

LEARNING OBJECTIVES

After you have finished studying this chapter, you should be able to:

1. Define the term energy and contrast potential and kinetic energy.

2. State the first and second laws of thermodynamics and discuss their applications to living organisms and to the ecosphere.

3. Distinguish between endergonic and exergonic reactions and explain how they may be coupled so that the second law of thermodynamics is not violated.

4. Describe the energy dynamics of a reaction that is in equilibrium.

5. Explain the function of enzymes and describe how they work.

6. Describe factors, such as pH and temperature, that influence enzymatic activity.

7. Compare the action and effects of the various types of enzyme inhibitors (e.g., competitive and noncompetitive inhibitors).

8. Describe the chemical structure of ATP and its role in cellular metabolism.

9. Describe the role in metabolism of hydrogen and electron acceptors, such as NAD.

1. In what part of the chapter will you probably get the information that will enable you to define "energy" and contrast potential and kinetic energy? Use the outline on page 105 to help you.

2. In what part of the chapter will you get information to aid you in accomplishing the second objective? the third?

3. How many objectives are there that deal with enzymes in some way? Which ones are they? (4, 5, 6, 7) Why do so many of the Learning Objectives deal with enzymes?

III. Read the introductory paragraph on page 106 and the summary on pages 119, 120, and 121. Take the Post-Test on page 121. Use your vocabulary cards as well as information from the introduction and summary to help you fill in the blanks.

IV. Vocabulary

1. Read through the vocabulary terms to find ones that "fit together."

   A. You should notice "energy" and "calorie" in more than one term. How do the terms that have the same word parts differ? How are they alike? What is the relationship between calorie and energy?

   B. There are two parts, "-gonic" and "-phosphate" which are repeated. What do these word parts mean? How are the terms in which they are found alike? How are they different?

   C. The term "competitive" is found in your vocabulary. What do you think it means? Why is it found in a chapter about energy?

   D. The word inhibition or inhibit is used in several vocabulary terms. What does it mean in this context?

V. Flip through the pages of Chapter 6 (106-119).

1. Note the relationship between the BOLD FACE headings and the Dark Face subheadings and the outline on page 105.

2. The figures in this chapter look different from those in Chapters 4 and 5. How are they different? Are they easier or harder to understand?

DURING READING ACTIVITIES

I. Before you start to read the section "WHAT IS ENERGY?" think of at least two questions you expect to find answers to when you read the section. Examples might be, what is the definition of energy? and how does energy work. As you read, look for the answers to these questions. When you finish the section, write down the answers if you found them.

II. Use the darker type as a guide to which terminology is important. Continue to circle those terms and any others you don't know. Be sure to underline the definitions and make vocabulary cards for any terms not already included in the vocabulary section.

III. Marking the text by first reading a whole paragraph and then underlining the main ideas and, finally, making notes in the margin should be easier now. Check the section from page 112 and see how your marking compares. Answer the following questions about the textbook marking.

1. Why are the asterisks (stars) used? Did you do something to remind yourself that this is an important concept?

2. Will you need to read the example again? If you understand the idea that intercellular reactions work together (as a team in which all members work together in a certain order) you should <u>not</u> need to reread this example.

3. Begin now to develop your own shorthand system of marking your text. Examples include asterisks (*1) for important ideas, arrows to show that one thing causes another (———>).

ENZYMES: CHEMICAL REGULATORS

Example of how intracellular reactions work

The thousands of <u>chemical</u> <u>reactions</u> that take place <u>within</u> <u>the</u> <u>living</u> <u>cell</u> <u>must</u> <u>be</u> <u>precisely</u> <u>controlled</u> <u>and</u> <u>coordinated</u>. These <u>reactions</u> <u>interlock</u> somewhat as a multitude of assembly processes interlock on a modern-day production line. In the manufacturing of an automobile, for example, it might not be possible to complete a car without some seemingly minor part, such as a ball bearing or special bolt. Very elaborate planning and scheduling are necessary to ensure that all parts are delivered to the plant by the time they are needed and that each worker performs a small part of the overall job at precisely the right time.

It is this <u>concept</u> <u>of</u> <u>control</u> that we wish to emphasize. Although energy can be released from glucose and other fuel molecules by burning, cells could not survive such high temperatures nor effectively utilize energy that is released in short sudden bursts.

Cells require a slow, steady release of energy that they must be able to regulate to meet metabolic energy requirements. Accordingly, fuel molecules are slowly oxidized and energy is extracted in small amounts during cellular respiration, which includes sequences of 30 or more reactions. In fact, most cellular metabolism proceeds by a series of steps so a given molecule may go through as many as 20 or 30 chemical transformations before it reaches some final state. And then the apparently completed molecule may be pre-empted by yet another chemical pathway so as to be totally transformed or consumed in the course of metabolism. The changing needs of the cell require a system of flexible chemical control. The key elements of this control system are the remarkable enzymes.

AFTER READING ACTIVITIES

I. Review your marking: whatever you underlined or highlighted and all of your margin notes. Was what you marked and wrote enough to remind you what the important facts and ideas in each section are? If not, be sure that you correct your marks.

II. Vocabulary

1. Did you categorize your terms before you began reading? How many of the terms do you already know? How many additional terms did you need to add.

2. Study the terms new to you in small groups of 5-7 terms

III. Answer the following questions.

Question

1. Every activity performed by a cell or living organism requires a continuous input of ___________.

2. It is important to be able to measure energy transformations. ___________ is the convenient form of energy to be measured. Furthermore, all other forms of energy can be converted into ___________.

3. Plants which carry out photosynthesis are organisms which transform light ___________ into ___________ energy stored in the ___________ bonds.

4. With time the total energy available to do work does ___________. This is because the useful forms of energy are continuously degraded to ___________.

5. According to the second law of thermodynamics, the ___________ of the universe is continuously increasing.

6. Are car engines more of less efficient than cells?

7. In cells, biological fuel molecules are ___________ oxidized and the energy is extracted in ___________ amounts during cellular respiration.

8. Enzymes increase the ____________ at which chemical reactions occur, and they do not cause a chemical reaction to occur that normally could not occur.

9. Enzymes are very <u>specific</u>. A single specific enzyme will, in many cases, catalyze only one chemical reaction, or a ____________ closely related chemical reactions.

10. Enzymes form temporary chemical complexes with their ____________. These ____________ then break up and release the product and the original ____________ ____________ for reutilization.

11. The ____________ ____________ of an enzyme is that area of the enzyme molecule in which the ____________ molecule(s) bind temporarily.

12. In the course of an enzyme catalyzed reaction, the ____________ of the enzyme molecule changes slightly.

13. Explain the induced fit model for enzyme activity.

14. A ____________ is a functional part of the ____________ site of an enzyme which is required for enzyme activity.

15. Many vitamins are required in that they act as ____________ for some enzyme catalyzed reactions.

16. Identify two mechanisms by which a cell can regulate the amount of a specific enzyme that is present in the cell.

A.

B.

17. Enzymes work under specifically defined conditions of temperature, pH, and sale concentration. These conditions are referred to as ____________ or ____________ conditions.

18. Distinguish between reversible and irreversible enzyme inhibition.

19. In competitive inhibition the ____________ and the ____________ compete for the ____________ site of the enzyme.

20. ____________ ____________ is when a control molecule binds to a region of the enzyme molecule which is not the active site, and this results in reduced enzyme activity.

21. Nerve gas is an example of ____________ ____________ of enzyme activity.

22. ATP is a nucleoside with three attached phosphate groups, ADP has __________ attached phosphate groups, and AMP has _________ attached phosphate group(s).

23. ____________ is the energy currency of the cell.

24. ATP has ____________ energy bond linking the last phosphate to the rest of the molecule.

25. Cells do not store large quantities of ____________. In fact, bacterial cells have only a one second supply of this material.

26. During oxidation of biological molecules, there is a loss of a ____________ atom from the compound.

Chapter 6 - Answers

1. Energy
2. Heat, heat
3. Energy, chemical, chemical
4. Decrease, heat
5. Entropy
6. Cells are 55 percent efficient and cars are only 17 percent efficient
7. Slowly, small
8. Rate
9. Specific, few
10. Substrates, complexes, enzyme molecule
11. Active site, substrate
12. Shape
13. See page 113-114 in the test for the answer.
14. Cofactor, active
15. Coenzymes
16. A. Regulate the gene which controls the amount of the enzyme which is synthesized.
    B. In some cases the enzymes are present in an inactive form. These enzymes possess an activator or allosteric site to which an activator molecule binds and changes the shape of the enzyme molecule making it now active.
17. Optima, optimal
18. Irreversible inhibitors permanently inactivate the enzyme while reversible inhibitors do not.
19. Inhibitor, substrate, active
20. Noncompetitive inhibition
21. Irreversible inhibition
22. Two, one
23. ATP
24. High
25. ATP
26. Hydrogen

ENERGY

POTENTIAL ENERGY

KINETIC ENERGY

KILOCALORIE

CALORIE

FIRST LAW OF
ERMODYNAMICS

SECOND LAW OF
THERMODYNAMICS

ENTROPY

FREE ENERGY

SPONTANEOUS REACTIONS

EXERGONIC REACTIONS

ENDERGONIC REACTIONS

STORED ENERGY. IT HAS THE CAPACITY TO DO WORK ACCORDING TO ITS POSITION OR STATE.

ABILITY TO CAUSE CHANGE; CAPACITY TO DO WORK. BIOLOGICAL WORK - THE WORK TO MAKE NEW COMPOUNDS, MECHANICAL WORK SUCH AS MUSCLE CONTRACTION, ELECTRO-CHEMICAL WORK SUCH AS ACCUMULATING A SUBSTANCE AGAINST A CONCENTRATION GRADIENT.

IS EQUAL TO A 1000 CALORIES AND IS THE UNIT USED TO MEASURE HEAT IN BIOLOGICAL SYSTEMS.

THE ENERGY OF MOTION.

THE LAW OF CONSERVATION OF ENERGY. DURING ORDINARY CHEMICAL AND PHYSICAL PROCESSES ENERGY CAN BE TRANSFERRED AND CHANGED FROM ONE FORM TO ANOTHER, BUT ENERGY CANNOT BE CREATED OR DESTROYED.

HEAT REQUIRED TO RAISE THE TEMPERATURE OF 1 GRAM OF WATER 1°, FROM 14.5° TO $15.5^{\circ}C$. FOOD CALORIES ARE REALLY KILOCALORIES.

THE ENERGY THAT HAS BECOME RANDOMIZED IN SYSTEM AND IS NO LONGER AVAILABLE TO DO WORK. MEASURE OF DISORDER OF THE SYSTEM. ENTROPHY OF THE UNIVERSE INCREASES.

THE DISORDER OF THE UNIVERSE IS CONTINUOUSLY INCREASING. DURING ANY PROCESS IN WHICH ENERGY IS RELEASED, SOME OF THE ENERGY IS DISSIPATED AS HEAT AND IS NOT AVAILABLE TO DO WORK.

REACTIONS THAT CAN OCCUR WITHOUT OUTSIDE INPUT BUT WHICH RELEASE FREE ENERGY AND CAN PERFORM WORK.

THE ENERGY AVAILABLE TO DO WORK UNDER CONDITIONS OF CONSTANT TEMPERATURE AND PRESSURE.

REACTIONS WHICH REQUIRE AN INPUT OF ENERGY AND ARE NOT SPONTANEOUS.

SAME AS SPONTANEOUS REACTIONS.

EQUILIBRIUM

ENZYMES

SUBSTRATE

ACTIVE SITE

INDUCED-FIT MODEL FOR ENZYME ACTIVITY

COFACTOR

COENZYME

INHIBITORS

COMPETITIVE INHIBITION

NONCOMPETITIVE INHIBITION

IRREVERSIBLE INHIBITORS

PHOTOSYNTHESIS

PROTEIN MOLECULES WHICH ACT AS ORGANIC CATALYSTS, AND WHICH ARE NOT CONSUMED DURING THE CHEMICAL REACTION. ENZYMES AFFECT THE RATE AT WHICH CHEMICAL REACTIONS OCCUR. ENZYMES GREATLY REDUCE THE ACTIVATION ENERGY BARRIER NECESSARY TO INITIATE A CHEMICAL REACTION.

THE RATE OF CHANGE IN ONE DIRECTION IS EXACTLY EQUAL TO THE RATE IN THE OPPOSITE DIRECTION.

A REGION OR REGIONS ON OR NEAR THE SURFACE OF THE ENZYME MOLECULE TO WHICH THE SUBSTRATE MOLECULE(S) TEMPORARILY BIND. SUBSTRATE THEN ALTERED.

THE CHEMICALS UPON WHICH AN ENZYME OPERATES.

THE NONPROTEIN COMPONENT OF AN ENZYME THAT IS A FUNCTIONAL PART OF THE ACTIVE SITE.

AN EXPLANATION FOR HOW AN ENZYME WORKS WHICH SUGGESTS ACTIVE SITES ARE NOT RIGID, AND WHEN SUBSTRATE BINDS TO THE ACTIVE SITE, IT INDUCES A CHANGE IN THE SHAPE OF THE ENZYME.

CHEMICAL AGENTS WHICH CAUSE AN ENZYME TO BE NONFUNCTIONAL. THESE INHIBITIONS MAY BE REVERSIBLE OR IRREVERSIBLE.

AN ORGANIC, NONPOLYPEPTIDE COMPOUND THAT SERVES AS A COFACTOR. MANY VITAMINS ACT AS COENZYMES.

THE INHIBITOR BINDS WITH THE ENZYME AT THE CONTROL SITE, WHICH IS DISTINCT FROM THE ENZYME'S ACTIVE SITE, AND THIS BINDING CHANGES THE SHAPE OF THE ENZYME MOLECULE.

ENZYME INHIBITOR MOLECULES COMPETE WITH SUBSTRATE MOLECULES FOR THE ACTIVE SITE OF THE ENZYME.

PROCESS OCCURRING IN PLANTS IN WHICH PLANT PRODUCES ATP, NADPH, AND SUGAR MOLECULES. DEPENDENT ON LIGHT.

SUBSTANCES SUCH AS POISONS OR ANTIBIOTICS WHICH BIND TO AN ENZYME AND PERMANENTLY INACTIVATE OR DESTROY THE ENZYME ACTIVITY.

ADENOSINE TRIPHOSPHATE
(ATP)

ADENOISINE DIPHOSPHATE
(ADP)

ADENOSINE MONOPHOSPHATE
(AMP)

PHOSPHORYLATION

NICOTINAMIDE ADENINE
DINUCLEOTIDE (NAD)

FLAVIN ADENINE
DINUCLEOTIDE (FAD)

ADENOSINE MOLECULE WITH TWO ATTACHED PHOSPHATES. PRODUCT OF REMOVAL OF ONE PHOSPHATE FROM ATP.

THE ENERGY CURRENCY OF THE CELL. THE MOLECULE OF ATP IS A NUCLEOTIDE (NUCLEOSIDE TRIPHOSPHATE). ATP HAS THREE PHOSPHATE GROUPS.

PROCESS BY WHICH A PHOSPHATE GROUP IS ADDED TO A MOLECULE.

ADENOSINE MOLECULE WITH ONE ATTACHED PHOSPHATE. PRODUCE OF REMOVAL OF TWO PHOSPHATES FROM ATP.

ANOTHER ELECTRON OR HYDROGEN ACCEPTOR FUNCTION SIMILAR TO NAD.

A COENZYME WHICH FUNCTIONS AS A HYDROGEN ACCEPTOR TO TEMPORARILY PACKAGE LARGE AMOUNTS OF FREE ENERGY.

CHAPTER 7

PHOTOSYNTHESIS: CAPTURING ENERGY

BEFORE READING ACTIVITIES

I. Before you begin to read Chapter 7, review the major concepts of Chapter 6.

1. What is energy? What are the two types of energy? How are they measured?
2. Explain the two laws of thermodynamics and discuss their applications to living organisms and to the ecosphere.
3. What is "free energy? And what are "coupled reactions"?
4. What is equilibrium?
5. Explain what enzymes are and why they are important.
6. Explain the function of ATP.
7. What are hydrogen and electron acceptors?

II. Getting an overview

1. Turn to page 111 of your textbook and read the outline about photosynthesis. Answer the following questions.

   1. What ONE word is used in many sections of the outline? Why do you think it is used so often? (Did you choose the word "light"?)

   2. How many types of reactions of photosynthesis are there? How are these reactions different?

   3. What is the most important major concept you need to learn about from this chapter? How can you tell? (You should have said, "the reactions of photosynthesis.)

2. Read the Learning Objectives listed below and answer the questions which follow it.

   LEARNING OBJECTIVES

   After you have read this chapter you should be able to:

   1. Write a summary reaction for photosynthesis, explaining the origin and fate of each substance involved.
   2. Describe the internal structure of a chloroplast.
   3. Summarize the events of the light-dependent reactions of photosynthesis, explaining the role of light in the activation of chlorophyll.
   4. Contrast cyclic and noncyclic photophosphoryiation.
   5. Describe how proton gradients form ATP according to the chemiosmotic theory.
   6. Summarize the events of the light-independent reactions of photosynthesis.
   7. Discuss the advantages of the C_4 pathway.

   1. In what part of the chapter will you probably get the information to help you write a summary reaction for photosynthesis?

2. How would you describe the internal structure of a chloroplast? Turn to page 126 of the text and try to use the figure to describe it.

3. Use Table 7-1 to test yourself on the principal reactions of photosynthesis. Cover the information in Columns 2, 3, 4 and use the information in the reaction series column to see if you can tell what the process is, what materials are needed, and what the end products are. (See Chapter Objectives 3 and 6.)

4. Compare Figure 7-7 (p. 130) with 7-8 (p. 131). How does noncyclic photophosphorylation compare with cyclic photophosphorylation? Can you create figures similar to these without looking back at the figures? Practice until you can do this.

III. Read the introductory paragraph on page 123 and the summary on page 135. Take the Post-Test on page 136. Remember to use your vocabulary cards and the information from the introduction and summary to help you fill in the blanks.

IV. Vocabulary

1. Read the vocabulary terms and sort them into groups that "fit together."

    1. How many terms have the word part "chloro-"?
    2. Which terms relate to the description of a chloroplast?
    3. How many terms use the word part "photo-"? What does it mean? How do the terms that use it differ from each other? How are they alike?

V. Flip through the pages of Chapter 7 (123-134). Note the relationship between the BOLD FACE headings and the Dark Face subheadings and the outline on page 122.

DURING READING ACTIVITIES

I. Before you start to read the section "LIGHT AND LIFE" think of at least two questions you expect to find answers to when you read the section. Examples might be, what is light and what is life? Where is light found? How are life and light related, what do life and light have to do with photosynthesis? As you read, look for the answers to these questions.

II. Use the darker type as a guide to important terminology. Circle dark type vocabulary terms and any other terms you don't know. Again, underline the definitions of terms. You may even want to write the definitions in the textbook margins. Check to see that you have vocabulary cards for these terms.

III. Continue to mark the text, paragraph by paragraph. When you have read and underlined a paragraph, write a summary of the paragraph in the margin. Be sure to summarize the paragraph using your OWN WORDS. Check the paragraph from page 123 to see if your marks and summarization is similar to the one below.

LIGHT AND LIFE

Since life on our planet depends upon light, it seems appropriate to discuss the nature of light and how it permits photosynthesis to occur. Light is a very small portion of a vast, continuous spectrum of radiation, the electromagnetic spectrum [Figure 7-1]. All radiations in this spectrum behave as though they travel in waves. At the end of the spectrum are gamma rays with very short wavelengths (measured in nanometers). A wavelength is the distance from one wave peak to the next. At the other end of the spectrum are low-frequency radio waves, with wavelengths so long that they are measured in kilometers

1. Plant depends on light	1. Did you have three important facts in the margin?
2. Light allows photosynthesis to occur	2. Did you circle "wavelength" and underline the definition?
3. Light - part of electromagnetic spectrum	3. Did you use some shorthand systemlike = to mean is?

AFTER READING ACTIVITIES

I. Review and correct your textbook marking. Be sure that your summarizing notes in the margins take sense and that your shorthand notation still makes sense to you.

II. Vocabulary
Be sure you practice the new vocabulary terms. Don't forget to review term from Chapter 6. Note the relationship between vocabulary terms from Chapters 6 and 7.

III. Answer the following questions.

Questions

1. Biological systems make use of _______________ fuels, although our technological society has not devised a means to use these fuels.

2. What two groups of organisms are able to obtain their energy from the sun?

3. How many billion tons of food are produced by the producer organisms?

4. How do consumer organisms obtain their energy?

5. Give an example of a stable form of chemical energy.

6. Light behaves as a ____________ and a ____________.

7. Describe how photons interact with atoms to raise the energy level of the affected atom.

8. What is the fate of the electron which is excited by the incoming photon?

9. Why are many plants green?

10. What are the main raw materials for photosynthesis?

11. Briefly describe what happens to the water molecule during photosynthesis.

12. Examine Table 7-1
   A. What are the three light-dependent reactions of photosynthesis?

      1. ____________________
      2. ____________________
      3. ____________________

   B. In what subcellular organelle do these reactions occur?

   C. In which organelle structures do the light-dependent reactions occur?

13. Plants release oxygen while carrying out photosynthesis. Where does this oxygen come from?

14. When the plant carries out the light-independent reactions, it expends a lot of chemical energy. What does the plant synthesize by using all this energy?

15. During cyclic photophosphorylation as the electrons are passed down the reaction pathway, there is the formation of chemical energy in the form of two molecules of __________________ and ____________________. There is also the release of __________.

16. In cyclic photophosphorylation, chemical energy in the form of one molecule of __________ is formed. However, no __________ is formed nor is any _______________ released.

17. ____________ ______________ is a protein complex which forms ATP as a result of the ____________ gradient across the thylakoid membranes.

18. The ____________ ____________ is a series of reactions in which atmospheric __________________ is incorporated into carbohydrate molecules.

Chapter 7 - Answers

1. Hydrogen
2. Plants and algae
3. 200 billion tons
4. These organisms obtain energy from the producer organisms they eat
5. Energy in coal and oil
6. Wave, particle
7. Quantum of light (photon) strikes an atom or molecule. Energy pushes the electron to a higher energy level, to an orbit farther from the nucleus
8. The electron is accepted by a reducing agent
9. Leaves reflect most of the green light that strikes them
10. Water (H_2O) and carbon dioxide
11. The water molecule is split by the energy of the sun. The oxygen is liberated, and the hydrogen is combined with CO_2 to produce carbohydrates. How this occurs is not known.
12. Examine Table 7-1
    A. 1. Photochemical reactions
       2. Electron transport
       3. Chemiosmosis
    B. Chloraplast
    C. Thylakoid membranes
13. The oxygen comes from the photolysis of water.
14. The plant synthesizes carbohydrates.
15. ATP, reduced NADP+, oxygen
16. ATP, NADPH, oxygen
17. ATP synthetase, proton
18. Calvin cycle, CO_2 (carbon dioxide)

PHOTOSYNTHESIS

WAVE LENGTH

PHOTONS

GROUND STATE

CHLOROPHYLL

THYLAKOIDS

CHLOROPLASTS

GRANA

STROMA

LIGHT-DEPENDENT REACTIONS

PHOTOCHEMICAL REACTIONS

ELECTRON TRANSPORT

REFERS TO THE VARIOUS FORMS OF ELECTROMAGNETIC RADIATION; THE DISTANCE FROM ONE WAVE TO THE NEXT.

THE PROCESS IN WHICH THE PLANTS AND ALGAE OBTAIN THEIR ENERGY FROM THE SUN.

THE LOWEST ENERGY STATE AN ATOM POSSESSES.

THE ENERGY PARTICLE OF LIGHT. ENERGY OF PHOTON DIFFERS.

TINY FLATTENED SACS WITHIN CELLS WHICH ARE ABLE TO CARRY OUT PHOTOSYNTHESIS.

THE MAIN PIGMENT USED BY PLANTS IN PHOTOSYNTHESIS. ABSORBS LIGHT IN THE BLUE, VIOLET, AND RED REGIONS.

STACKS OF THYLAKOIDS. EACH GRANUM LOOKS LIKE A STACK OF COINS. (THYLAKOIDS)

STRUCTURES WITHIN THE PLANT CELLS WHICH CONTAIN THE THYLAKOIDS.

THYLAKOID MEMBRANE REACTIONS IN WHICH SUNLIGHT ENERGY IS USED TO SPLIT H_2O AND SYNTHESIZE ATP (ENERGY), ALSO REDUCES NADP.

THE FLUID-FILLED REGION OF THE CHLOROPLAST WHICH CONTAINS THE ENZYMES REQUIRED FOR PHOTOSYNTHESIS.

THE SECOND OF THE LIGHT DEPENDENT REACTIONS IN WHICH ELECTRONS ARE TRANSPOSED ALONG A CHAIN OF THYLAKOID MEMBRANE ACCEPTORS. RESULTS IN REDUCTION OF NADP AND THE SPLITTING OF H_2O TO CAUSE H+ BUILD UP IN THYLAKOIDS.

ONE OF THE LIGHT DEPENDENT REACTIONS IN WHICH CHLOROPHYLL IS ENERGIZED, AND AN ENERGIZED ELECTRON IS GIVEN TO AN ELECTRON ACCEPTOR.

CHEMIOSMOSIS

LIGHT-INDEPENDENT REACTIONS

PHOTOLYSIS

PHOTOSYSTEMS

REACTION CENTER

NONCYCLIC PHOTOPHOSPHORYLATION

CYCLIC PHOTOPHOSPHORYLATION

ATP SYNTHETASE

CALVIN CYCLE

CO_2 FIXATION

REACTIONS WHICH OCCUR IN THE STROMA OF THE CHLOROPLAST, AND RESULT IN THE FIXATION OF CARBON DIOXIDE TO FORM ORGANIC SUGAR MOLECULES AND OTHER CARBOHYDRATES.

THE THIRD OF THE LIGHT-DEPENDENT REACTIONS. THE PROTONS (H+) MOVE ACROSS THE THYLAKOID MEMBRANE DOWN A PROTON GRADIENT. THIS RESULTS IN THE NET PRODUCTION OF ATP.

CHLOROPHYLL MOLECULES AND ASSOCIATED ELECTRON ACCEPTORS ARE ORGANIZED INTO UNITS OF 400 CHLOROPHYLL MOLECULES.

THE SPLITTING OF WATER DURING PHOTOSYNTHESIS LEADING TO THE RELEASE OF OXYGEN TO THE ATMOSPHERE. THE H^+ FROM THE WATER IS USED TO FORM CHEMICAL ENERGY.

THE SET OF REACTIONS IN WHICH ELECTRONS FLOW THROUGH A SERIES OF ELECTRON ACCEPTORS WHICH IS INVOLVED IN ATP SYNTHESIS.

A SPECIAL PIGMENT MOLECULE OF A PHOTOSYSTEM WHICH GIVES UP ITS ENERGIZED ELECTRON TO AN ACCEPTOR MOLECULE.

PROTEIN COMPLEX EMBEDDED IN THYLAKOID MEMBRANE WHICH FORMS ATP.

ELECTRONS ORIGINATE FROM PHOTOSYSTEM I AND ARE RETURNED FROM THAT SAME PHOTOSYSTEM. TWO ELECTRONS ENTER PATHWAY AND PRODUCE ONE ATP. NO NADPH IS FORMED, NO OXYGEN RELEASED.

THE INCORPORATION OF ATMOSPHERIC CO_2 INTO CARBOHYDRATE MOLECULES.

THE REACTIONS OF THE LIGHT-INDEPENDENT CYCLE IN WHICH THE ENERGY OF NADPH AND ATP IS USED TO MANUFACTURE CARBOHYDRATES. CARBON DIOXIDE IS FIXED IN THIS PROCESS.

CHAPTER 8

ENERGY-RELEASING PATHWAYS: CELLULAR RESPIRATION AND FERMENTATION

BEFORE READING ACTIVITIES

I. Before you begin to read Chapter 8, review the concepts in Chapter 7.

1. What are major ideas you learned about photosynthesis?

   (You should have: relationship between light and life, absorbing light (or) chlorophyll), and the reactions of photosynthesis.)

2. What are the two types of photosynthetic reactions?

3. Summarize the events of each type of photosynthetic reaction.
   Hint: Be sure you include the photosystems, cyclic photophosphorylation, chemiosmosis, carbon fixation, the C_4 pathway.
   Write your answer out on another sheet of paper and then check it in the text.

4. What do you think is the relationship between photosynthesis and cellular respiration and fermentation?
   Hint: Look at the chapter titles. Photosynthesis is capturing energy and cellular respiration and fermentation are energy-releasing pathways.

II. Getting an overview

1. Turn to page 137 of your textbook and read the outline about energy-releasing pathways. Answer the following questions.

   1. What are the four reactions of cellular respiration?

   2. What do you think is the difference between the reactions of cellular respiration and the regulation of cellular respiration.

   3. What do you think anaerobic pathways are?

2. Read the Learning Objectives below and answer the questions which follow.

   1. On what page of the chapter will you probably get the information to accomplish each objective? Write the page numbers in the space next to the objective.

2. Note that the two types of activities expected of you are summarizing and comparing. What do these two terms mean to you?
3. Did you realize that Objectives 2, 4, and 5 all mention ATP? Be sure you review what you know about ATP from Chapters 6 and 7.

LEARNING OBJECTIVES

After you have read this chapter, you should be able to:

1. Write a summary reaction for cellular respiration, giving the origin and fate of each substance involved.
2. Summarize the events of glycolysis, giving the key organic compounds formed and the number of carbon atoms in each; indicate the number of ATP molecules used and produced and the transactions in which hydrogen loss occurs.
3. Summarize the events of the citric acid cycle, beginning with the conversion of pyruvic acid to acetyl-CoA; indicate the fate of carbon-oxygen segments and of hydrogens removed from the fuel molecule.
4. Summarize the operation of the electron transport system, including the reactions by which a gradient of protons is established across the inner mitochondrial membrane; explain how the proton gradient drives ATP synthesis.
5. Compare aerobic respiration with anaerobic pathways in terms of ATP formation, final hydrogen acceptor, and end products; give two specific examples of anaerobic pathways.

3. Be sure to read the introduction and summary. Take the Post-Test.

4. Read through the vocabulary terms and try to determine their meanings from the word parts you already know. Note that there are not very many new vocabulary terms.

DURING READING ACTIVITIES

I. Remember that you will be expected to SUMMARIZE the information in the chapter. Since a summary RETELLS the information in a SHORTER form, you need to think about how you can get the main ideas about each topic and restate so someone else will understand it.

II. Be sure to mark your text. After you read and mark a group of paragraphs (reading from one sub-heading to another), write a summary of what you learned. Your first summary should be after you complete reading page 139. Your second summary should be after you finish reading the section on Glycolysis on page 140.

AFTER READING ACTIVITIES

I. Review and correct your textbook marking.

II. Check your summaries. Do they include the main ideas? Did you do a summary for each of the concepts listed in the objectives (1,2,3,4)? If not, go back and summarize each of these.

III. Test yourself on the new vocabulary. Be sure to review all vocabulary from Chapters 6 and 7. Remember that Chapters 6, 7 and 8 are a unit of the text which go together to explain the concept of the "Life and Flow of Energy."

IV. Answer the following questions.

1. List the four phases of the stepwise oxidation of glucose.

   A. ______________________________
   B. ______________________________
   C. ______________________________
   D. ______________________________

2. The reactions of the four phases of cellular respiration occur in different cellular compartments. Glycolysis occurs in the ________________ while the reactions of the citric acid cycle occur in the _______________.

3. Each reaction of glycolytic pathway is regulated by a particular, specific _______________.

4. Glycolysis is divided into two phases. In the first phase the six carbon glucose is broken into two _____________ carbon glyceraldehyde-3-phosphate molecules. The second phase, two ____________ atoms are removed and the molecules are rearranged to form _______________.

5. Glycolysis yields only a small amount of energy, for ________ molecules of ATP are formed for each molecule of glucose which is metabolized to pyruvate.

6. ATP molecules are ________________ in the first phase of glycolysis and _______________ in the second phase.

7. What is the ultimate fate of the carbon atom removed from pyruvate when it is converted to acetyl coenzyme A?

8. ___________________ forms from the combination of the _________ ____________ and coenzyme A.

9. Fat molecules can be broken down to form ______________________ which in turn can enter the respiratory (metabolic) pathway at this point.

10. Consult the Focus on -- The Citric and Cycle on page 145. Starting with citrate, which is formed by the combination of acetyl coenzyme A and oxaloacetate, list the names of the compounds of the citric acid cycle in proper sequence.

11. In the course of the citric acid cycle, two molecules of carbon are split off the fuel molecules. These carbons are released as _______________.

12. During the citric acid cycle, hydrogen atoms are removed from various molecules and transferred to the hydrogen acceptor molecules ___________ and ___________. The energy in these hydrogen atoms is used to make ________.

13. The hydrogens are transferred from one to another of the fifteen different electron acceptor molecules. These mitochondrial electron acceptor molecules are localized in the __________ ______________ ______________.

14. Human beings require oxygen to survive. What is the major functions of this oxygen?

15. Examine Figure 8-5 (p. 147).

    A. Where do the protons which are pumped across the inner mitochondrial membrane accumulate?

    B. These protons then move back across the inner mitochondrial membrane. What occurs when this happens?

    C. Describe the cristae.

    D. Where in the mitochondria do the enzymes of the citric acid cycle localize?

16. Examine Figure 8-2 (page 139).

    A. How many carbons in an acetyl COA molecule?

    B. How many carbon atoms in an oxaloacetate molecule?

    C. How many carbon atoms in citrate?

    D. Since the citric acid cycle starts with citrate and ends with oxaloacetate, what happens to the two carbon atoms which are removed in the intervening reactions?

17. Continue to examine Figure 8-2 (page 139).

    A. In how many reactions of the citric acic cycle are hydrogen atoms removed?

B. These hydrogen atoms are then transferred to the hydrogen carrier molecules ___________ and ___________ which in turn carry the hydrogen atoms to the electron transport system. The transport of electrons down the electron transport system establishes the ___________ ___________ across the inner mitochondrial membrane. The movement of these ___________ back across the membrane draws the synthesis of the high energy ___________ molecules.

18. Examine Table 8-2.

A. How many substrate level ATP molecules are formed during the complete oxidation of a glucose molecule?

B. Bacteria which live anaerobically generate far less ATP than organisms which can utilize oxygen. How many ATP molecules can an anaerobic bacterium get from the oxidation of a glucose molecule?

C. Aerobic organisms form far more ATP from the oxidation of glucose than do anaerobic organisms. How many more times ATP is formed aerobically than anaerobically?

19. Examine Figure 8-6 (page 150)

When oxygen is present, the hydrogen atoms removed from the sugar molecules during glycolysis pass through the electron transport system. When oxygen is absent, these hydrogens are disposed of in different ways. List some of these other ways.

A.

B.

C.

20. What accounts for the souring of milk or fermenting of cabbage to form sauerkraut?

Chapter 8 - Answers

1. A. Glycolysis
   B. Formation of acetyl coenzyme A
   C. The citric acid cycle
   D. Electron transport and the chemiosmotic formation of ATP.
2. Cytoplasm, mitochondria
3. Enzyme
4. Three, hydrogen, pyruvate
5. Two
6. Consumed, produced
7. The carbon is removed with an oxygen to form CO_2, and this CO_2 is expelled by the lungs.
8. Acetyl coenzyme A, the acetyl group
9. Acetyl coenzyme A
10. Citrate, cis-aconitate, Isocitnate, x-Ketoglutarate, Succnyl CoA, Succinate, Fumate, Malate, Oxaloacetate.
11. Carbon dioxide
12. FAD, NAD, ATP
13. Inner, mitochondrial membrane
14. Oxygen is the final electron acceptor, so that electrons from glycolysis and the citric acid cycle are combined with oxygen and H+ to form water.

15. A. Between the inner and outer mitochondrial membranes
    B. ATP is formed
    C. The cristae are formed by the extensive folding of the inner mitochondrial membrane.
    D. The enzymes of the citric acid cycle are localized in the matrix portion of the mitochondrion, which is bounded by the inner mitochondrial membrane.
16. A. Two
    B. Four
    C. Six
    D. The two carbons are removed and given up as CO_2.
17. A. Four
    B. NAD, FAD, proton gradient, protons, ATP
18. A. Two
    B. Only two
    C. 36 to 2 or 38 to 2, or 18 to 19 times more ATP formed by aerobic organisms
19. A. Nitrate serves as a final electron acdeptor
    B. Yeast split CO_2 off of pyrurate and add hydrogen to the resulting molecule to form ethyl alcohol.
    C. Hydrogens can be transferred to a pyrurate molecule to form lactate.
20. Bacteria ferment sugar molecules and form lactate.

CELLULAR RESPIRATION

GLYCOLYSIS

FORMATION OF ACETYL COENZYME A

CITRIC ACID CYCLE A
(KREBS CYCLE)

ELECTRON TRANSPORT

ACETYL COENZYME A

OXIDATIVE PHOSPHORYLATION

ANAEROBIC METABOLISM

AEROBIC METABOLISM

SERIES OF TEN REACTIONS DURING WHICH GLUCOSE IS DEGRADED TO PYRUVATE; YIELDS 2 ATPs. HYDROGENS RELEASED. DOES NOT REQUIRE OXYGEN. OCCURS IN CELL CYTOPLASM.

PROCESS BY WHICH LIVING CELLS BREAK DOWN FUEL MOLECULES TO CARBON DIOXIDE (CO_2) AND WATER (H_2O) WITH THE RELEASE OF ENERGY STORED IN CHEMICAL BONDS.

SERIES OF REACTIONS WHICH OCCUR IN THE MITOCHONDRIA IN WHICH A FUEL MOLECULE IS DEGRADED TO HYDROGEN AND CO_2; REQUIRES OXYGEN.

IN THE MITOCHONDRIA PYRUVATE IS DEGRADED AND COMBINED WITH COENZYME A TO FORM ACETYL COENZYME A. CO_2 IS RELEASED.

A LARGE, COMPLEX MOLECULE WHICH FORMS FROM THE COMBINATION OF THE TWO CARBON ACETYL GROUP WITH COENZYME A.

SERIES OF SEVERAL REACTIONS WHICH OCCUR IN THE MITOCHONDRIA. HYDROGEN ATOMS OR ELECTRONS ARE PASSED ALONG A SERIES OF MOLECULES. ENERGY RELEASED ESTABLISHES PROTON GRADIENT ACROSS MITOCHONDRIAL MEMBRANE WHICH IN TURN DRIVES ATP SYNTHESIS. REQUIRES OXYGEN.

THE METABOLISM OF GLUCOSE IN THE ABSENCE OF OXYGEN. GENERALLY REFERS TO GLYCOLYSIS.

THE ENTIRE PROCESS WHICH INVOLVES THE FLOW OF ELECTRONS DOWN THE ELECTRON TRANSPORT SYSTEM, THE ESTABLISHMENT OF THE PROTON GRADIENT, AND THE PHOSPHORYLATION OF ADP TO FORM ATP.

THE METABOLISM OF GLUCOSE OR OTHER SUGAR MOLECULES IN THE PRESENCE OF OXYGEN. INCLUDES GLYCOLYSIS, THE CITRIC ACID CYCLE, AND OXIDATIVE PHOSPHORYLATION.

CHAPTER 9

CHROMOSOMES, MITOSIS, AND MEIOSIS

BEFORE READING ACTIVITIES

I. Before you begin to read, note that Chapter 9 is the first chapter in a new part (Part III) of the text which discusses the "CONTINUITY OF LIFE: CELL DIVISION AND GENETICS. The authors suggest that there are five major concepts which should be understood:

CELL DIVISION AND GENETICS

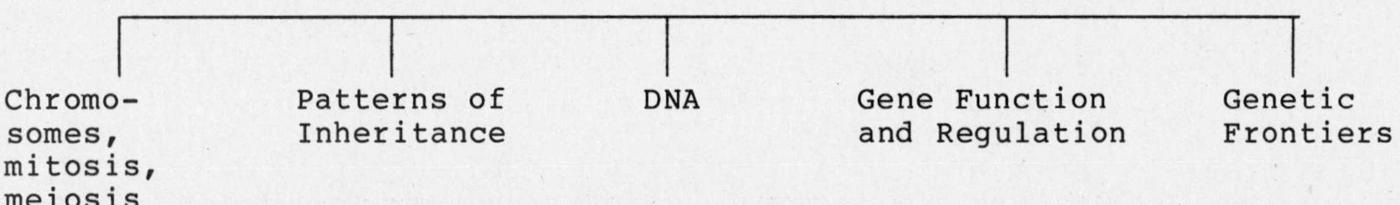

Each of these concepts is of equal importance (because they are full chapters) and each explains something about a single topic (cell division and genetics) and; because they are all about a single topic, they will have many things in common: vocabulary, concepts, use of information learned in one chapter repeated in another.

II. Getting an Overview

1. Turn to page 155. Read the outline about chromosomes, mitosis, and meiosis.

   1. What are the major ideas discussed in the chapter?

      1. ____________________ 4. ____________________
      2. ____________________ 5. ____________________
      3. ____________________

2. Which is a broader concept, mitosis or prophase? How can you tell by reading the outline?

3. Which is a broader concept, the second meiotic division or synapsis, linkage and crossing-over?

4. Can you construct a "picture" of the structure of the chapter?

CHROMOSOMES, MITOSIS, AND MEIOSIS

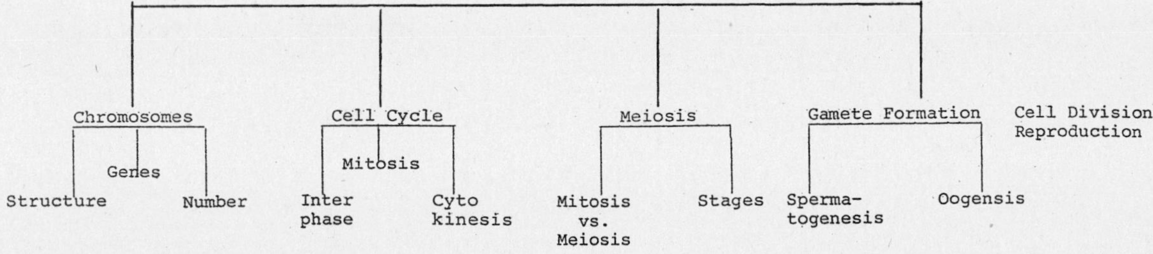

A picture like this is called a graphic organizer. A graphic organizer helps you to understand the way in which concepts fit together and it helps you to remember which ideas "go together."

III. Read the introductory paragraph on page 156 and the summary on pages 173 and 174. Take the Post-Test on page 175. Be sure to use the vocabulary cards and the information from the graphic organizer, outline, introduction, and summary to help you.

IV. Read the Learning Objectives on page 155.

1. Decide which pages of the chapter will give you information to enable you to achieve each objective. (Your answer for the first objective should be pages 156 and 157.)

2. Notice the terms used: describe, distinguish between, and contrast. What do these terms ask you to do?

V. Vocabulary

1. Read through the terms and their definitions. Sort them into groups that "fit" together.

2. How many terms are there which use the word "phase"? How do these terms relate to each other?

3. There are several terms which use the words "first" and "second." How do these terms relate to each other?

4. There are some terms which seem to be common language. What do they mean in biology? (linked, crossing over, recombination, lampbrush, asters, maternal, paternal, tetrad)

DURING READING ACTIVITIES

I. Remember that you will be expected to DESCRIBE the various concepts in this chapter (structure of a chromosome, main events of cell cycle stages, etc.) Describe means RETELL information giving the IMPORTANT POINTS of each. This is similar to summarize. Steps to use while you read should include:

1. Mark the main ideas of each paragraph.
2. Mark the <u>most</u> <u>important</u> <u>details</u>. Use some system to differentiate between main ideas and important details. You might asterisk main ideas or underline main ideas twice or use a curved line under main ideas.

3. When you have finished reading from one subheading to another, write your summary or description of the concept presented. To describe the structure of a mature sperm (Objective 7), you should first read pages 169 and 170. Be sure you read Figure 9-14. As you read, mark each paragraph. Then write your description of the mature sperm in the margin on page 170 or on a separate sheet of paper.

II. You also need to be able to "DISTINGUISH BETWEEN" or "CONTRAST" ideas. To do this, you must be able to tell how two (or more) things are DIFFERENT. How does "haploid" differ from "diploid"? How are the events of mitosis different from meiosis?

AFTER READING ACTIVITIES

I. Review your textbook marking. If you have not marked enough to make sense when you review, be sure to correct this. If you have marked too much, correct this by using a second type of marking. If you highlighted, go back and underline only the important parts. If you underlined, go back and highlight only the important parts.

II. Test yourself on the new vocabulary. In this chapter there are alot of new terms to learn. To remember these new terms, you need to get them stored in your long-term memory.

1. One way to do this is to sort the information into categories (groups that make sense). In biology you can sort these terms into those that are used to describe or explain a particular idea or concept. For example, you should think of transformation, conjugation, and transduction together because they are all three mechanisms of genetic recombination.

2. A second thing you should do to help retain this information is to relate it to something you already know. For example, what do you already know about the terms transformation and conjugation? Is what you already knew the same as what you need to know now? What can you use to help you remember the new meanings of these terms?

III. Go back to the graphic organizer at the beginning of this chapter. Try to explain to yourself what concepts are involved in each of the headings and subheadings. Fill in as many details as you can.

IV. Answer the following questions.

1. Examine Figure 9-2.

   A. Describe the arrangement of the DNA and the histone molecules.

2. Are chromosomes uniform in size and shape?

3. Approximately how many genes are contained in a single chromosome?

4. Write out the definition of a gene.

5. How many different genes are there in a typical mammalian cell?

6. Every species has a ____________ or ____________ number of chromosomes in each of its somatic cells.

7. Humans have ____________ chromosomes in each somatic cell. However, these chromosomes are present as similar pairs, so humans have 23 pairs of ______________ ______________.

8. In humans 46 is the ____________ number of chromosomes and 23 is the ___________ number of chromosomes.

9. Cell division consists of two separate processes. These are ___________ and ____________.

10. Examine Figure 9-4 showing the cell cycle.

    A. Most of the life of the cell is spent in interphase. What are the three subphases of interphase?

    B. Aside from interphase, what is the other phase of the cell cycle?

    C. What are the four stages of mitosis?

11. Examine Figure 9-6.

    A. In late prophase, how many chromatids in each of the sex chromosomes?

    B. What is the structure joining the chromotids called?

    C. In anaphase, the chromosomes are pulled to opposite poles of the cell. How many chromatids per chromosome?

12. Which stage of mitosis is chosen to examine chromosomes in the light microscope?

13. Match the stage of mitosis with the description of what is occurring.

A. Prophase	1. Chromosomes are lined up along the equatorial plane.
B. Metaphase	2. Final stage of mitosis.
C. Anaphase	3. Stage when centromeres split and sister chromotids pulled to opposite poles of the cell.
D. Telophase	4. Chromatin condenses, nuclear membrane breaks down.

14. What are the two processes which occur during cell division?

15. Name the cells which are capable of undergoing meiosis.

16. Meiosis and mitosis are very similar, but there are some important differences between these two processes. List these differences.

17. Humans have a total of 46 chromosomes. Of these, how many are maternal and how many are paternal chromosomes?

18. What is the consequence of crossing over?

19. In meiosis there is ____________________ of chromotids between meiosis I and meiosis II.

20. Meiosis I is characterized by the ____________________ of each homologous pair of chromosomes, so that one member of each pair ends up in each daughter cell. Each chromosome consists of ____________ chromatids.

21. Meiosis II is very similar to another process, ____________.

22. Upon completion of meiosis, each daughter cell contains a ____________ number of chromosomes, and each chromosome consists of ____________ chromatid(s).

23. Examine Figure 9-12. At which stage of meiosis do each of the following occur.

    A. Homologous chromosomes separate

    B. Chromosomes (DNA) duplicated

    C. Four gametes produced, each haploid

    D. Chromatids line up along equatonal plane

    E. Chromatids separate, move to opposite poles

24. Name the three parts of a human sperm.

25. ____________ is the formation of ova or eggs.

Chapter 9 - Answers

1. The DNA is wrapped around the outside of a cluster of eight molecules of histone.
2. No, they are quite diverse.
3. Hundreds or even thousands of genes per chromosome.
4. See vocabulary card.
5. Approximately 25,000 to 50,000
6. Characteristic, specific
7. 46, homologous chromosomes
8. Diploid, haploid
9. Mitosis, cytokinesis
10. A. G_1 or first gap phase, synthesis phase or S phase, and G_2 or second gap phase.
    B. Mitosis
    C. Prophase, metaphase, anaphase, and telophase
11. A. Two
    B. Centromere
    C. One
12. Metaphase, the chromosomes are completely condensed.
13. A. 4
    B. 1
    C. 3
    D. 2
14. Mitosis and cytokinesis
15. Gametes (eggs and sperm) and spores
16. See pp. 165-166 and Figure 9-9
17. 23 maternal chromosomes and 23 paternal chromosomes
18. Because of the reciprocal exchange of genetic material, genes can be assorted and new combinations formed. Increases genetic variability.
19. Duplication
20. Separation, two
21. Mitosis
22. Haploid, one
23. A. Anaphase I
    B. Interphase prior to Meiosis I
    C. Telophase II
    D. Metaphase II
    E. Anaphase II
24. Head, midpiece, tail
25. Oogenesis

HEREDITY

GENETICS

HISTONES

NUCLEOSOME

CENTROMERE

GENE

HOMOLOGOUS CHROMOSOMES

DIPLOID (2N)

MEIOSIS

HAPLOID

CHROMOSOMES

CELL CYCLE

THE STUDY OF HEREDITY.

THE TRANSFER OF BIOLOGICAL (GENETIC) INFORMATION FROM PARENT TO OFFSPRING.

THE REPEATING UNITS OF DNA WOUND AROUND A CLUSTER OF EIGHT MOLECULES OF HISTONE.

BASIC PROTEINS (POSITIVELY CHARGED) WHICH ASSOCIATE WITH THE DNA IN A SPECIFIC WAY.

SEQUENCE OF DNA THAT CONTAINS THE INFORMATION TO SYNTHESIZE A SPECIFIC RNA MOLECULE. MANY RNA MOLECULES DIRECT THE SYNTHESIS OF SPECIFIC PROTEINS.

SMALL, LIGHT-STAINING, CIRCULAR ZONE AT A SPECIFIC LENGTH ALONG THE LENGTH OF THE CHROMOSOME. CONTROLS THE MOVEMENT OF THE CHROMOSOME DURING CELL DIVISION.

SITUATION WHEN A CELL HAS TWO MEMBERS OF EACH AND EVERY PAIR OF CHROMOSOMES.

TWO CHROMOSOMES WHICH ARE VERY SIMILAR TO ONE ANOTHER AND CARRY INFORMATION GOVERNING THE SAME TRAITS. THIS INFORMATION IS NOT NECESSARILY THE SAME.

SITUATION WHEN A CELL CONTAINS ONE MEMBER OF EACH CHROMOSOME PAIR.

A SPECIAL TYPE OF CELL DIVISION WHICH OCCURS ONLY IN EGG AND SPERM CELLS, AND WHICH RESULTS IN A REDUCTION OF THE NUMBER OF CHROMOSOMES IN EACH CELL TO ONE CHROMOSOME OF EACH HOMOLOGOUS PAIR.

THE PERIOD FROM BEGINNING OF ONE CELL DIVISION TO THE BEGINNING OF THE NEXT CELL DIVISION.

PACKAGES OF STORED GENETIC INFORMATION.

CYTOKINESIS

MITOSIS

INTERPHASE

G_1 - FIRST GAP PHASE

SYNTHESIS PHASE (S PHASE)

G_2 - SECOND GAP PHASE

PROPHASE

CHROMATIDS

ASTERS

MITOTIC SPINDLE

METAPHASE

ANAPHASE

A COMPLEX DIVISION WHICH PRECISELY DISTRIBUTES A COMPLETE SET OF NEWLY DUPLICATED CHROMOSOMES TO EACH DAUGHTER NUCLEUS. ENSURES THAT EACH NEW CELL CONTAINS THE IDENTICAL NUMBER AND TYPE OF CHROMOSOMES PRESENT IN THE ORIGINAL CELL. CONSISTS OF STAGES: PROPHASE, METAPHASE, ANAPHASE AND TELOPHASE.

THE ACTUAL DIVISION OF THE CYTOPLASM TO YIELD TWO DAUGHTER CELLS. STARTS IN LATE ANAPHASE AND IS COMPLETED IN TELOPHASE.

TIME BETWEEN MITOSIS AND BEGINNING OF DNA REPLICATION. CELL PREPARES FOR THE DUPLICATION OF ITS DNA.

PHASE IN WHICH CELL SPENDS MOST OF ITS LIFE, SYNTHESIZES MATERIALS FOR GROWTH AND MAINTENANCE. CHROMOSOME REPLICATION OCCURS.

PHASE FOLLOWING COMPLETION OF THE S PHASE. ORGANELLE MATERIALS SYNTHESIZED. INCREASED PROTEIN SYNTHESIS.

DNA IS SYNTHESIZED DURING THE S PHASE. AMOUNT OF DNA IN CELL DOUBLES AS A RESULT OF DNA SYNTHESIS.

DUPLICATED CHROMOSOMES JOINED AT THEIR CENTROMERE. IN PROPHASE EACH CHROMOSOME CONSISTS OF TWO CHROMATIDS.

FIRST STAGE OF MITOSIS, CHROMATIN CONDENSERS. FORMS CHROMOSOMES WHICH ARE SHORT AND THICK. CONSISTS OF TWO CHROMATIDS PER CHROMOSOME. NUCLEAR ENVELOPE BREAKS DOWN.

CLUSTER OF MICROTUBULES EXTENDING FROM CENTRIOLE REGION.

CLUSTERS OF MICROTUBULES. INVOLVED IN CELL DIVISION.

STAGE WHEN THE CENTROMERES OF THE SISTER CHROMATIDS ARE SPLIT APART. EACH CHROMATID IS AN INDIVIDUAL CHROMOSOME. SEPARATED CHROMOSOMES MOVE TO OPPOSITE POLES OF THE CELL.

SECOND STAGE OF MITOSIS. SHORT PERIOD DURING WHICH TIME THE CHROMATIDS ARE LINED UP ALONG THE EQUATORIAL PLANE. MITOTIC SPINDLE FORMED.

TELOPHASE

CHROMATIN

CYTOKINESIS

MEIOSIS

FIRST MEIOTIC DIVISION

SECOND MEIOTIC DIVISION

SYNAPSIS

MATERNAL CHROMOSOME

PATERNAL CHROMOSOME

TETRAD

LINKED

CROSSING OVER

EXCEPT FOR DURING MITOSIS AND MEIOSIS, THE NUCLEAR MATERIAL IS NOT ORGANIZED INTO CHROMOSOMES. THIS MATERIAL IS DISPERSED. IT CONSISTS OF 60% PROTEIN, 35% DNA, AND 5% RNA.

FINAL STAGE OF MITOSIS. RETURN TO INTERPHASE CONDITIONS. CHROMOSOMES UNCOIL. NEW NUCLEAR MEMBRANE FORMS.

SPECIAL CELL DIVISION FOR EGGS, SPERM AND SPORES. CONSISTS OF TWO CELL DIVISIONS - MEIOSIS I AND MEIOSIS II. EACH DIVISION CONSISTS OF FOUR STAGES SIMILAR TO MITOSIS.

THE DIVISION OF THE CYTOPLASM TO YIELD TWO DAUGHTER CELLS. THIS PROCESS USUALLY ACCOMPANIES MITOSIS. BEGINS AT LATE ANAPHASE AND EXTENDS THROUGH TELOPHASE.

SEPARATES THE CHROMATIDS INTO INDIVIDUAL CHROMOSOMES CONSISTING OF ONE CHROMATID PER CHROMOSOME.

SEPARATES THE MEMBERS OF EACH HOMOLOGOUS PAIR OF CHROMOSOMES AND PARCELS THEM OUT TO DAUGHTER CELLS. EACH CHROMOSOME IS DUPLICATED--TWO CHROMATIDS PER CHROMOSOME.

THE CHROMOSOME OF A HOMOLOGOUS PAIR CONTRIBUTED BY THE ORGANISMS MOTHER.

THE PAIRING UP OF HOMOLOGOUS CHROMOSOMES SO AS TO LINE UP ALONG THEIR ENTIRE LENGTH.

A COMPLEX FORMED WHEN THE TWO CHROMOSOMES OF A HOMOLOGOUS PAIR UNDERGO SYNAPSIS. EACH CHROMOSOME WAS DUPLICATED AND CONSISTS OF TWO CHROMATIDS, THUS A HOMOLOGOUS PAIR CONSISTS OF FOUR CHROMATIDS.

THE CHROMOSOME OF A HOMOLOGOUS PAIR CONTRIBUTED BY THE ORGANISM'S MOTHER.

DURING SYNAPSIS HOMOLOGOUS CHROMOSOMES BECOME ENTANGLED, AND GENETIC MATERIAL CAN BE EXCHANGED BETWEEN HOMOLOGOUS CHROMOSOMES, MATERNAL AND PATERNAL CHROMATIDS BREAK AND REJOIN RECIPROCALLY.

ALL GENES ON SINGLE CHROMOSOME EXIST ON A COMMON STRUCTURE.

GENETIC RECOMBINATION

SPORANGIA

SPORES

GONADS

SPERMATOGENSIS

TESTES

SEMINIFEROUS TUBULES

PRIMARY SPERMATOCYTE
SECONDARY SPERMATOCYTE
SPERMATID

ACROSOME

OOGENIS

OVARY

OVULATION

PLANT STRUCTURES IN WHICH MEIOSIS OCCURS.

THE EXCHANGE OF GENETIC INFORMATION BETWEEN HOMOLOGOUS CHROMOSOMES.

SPECIALIZED REPRODUCTIVE STRUCTURES IN ANIMALS IN WHICH MEIOSIS OCCURS.

PLANT REPRODUCTIVE CELLS.

MALE GONAD.

SPERM PRODUCTION IN ANIMALS.

STAGES IN FORMATION OF MATURE SPERM.

TUBULES WITHIN TESTES IN WHICH SPERMATOGENESIS OCCURS.

THE PRODUCTION OF EGGS OR OVA.

STRUCTURE IN FRONT PART OF SPERM THAT HELPS THE SPERM PENETRATE THE EGG.

THE RELEASE OF A MATURE OOCYTE FROM THE OVARY.

THE FEMALE GONAD.

PLASMIDS

TRANSVERSE BINARY FISSION

TRANSFORMATION

MEANS BY WHICH BACTERIA REPRODUCE. CELL DEVELOPS TRANSVERSE CELL WALL AND THEN DIVIDES INTO TWO DAUGHTER CELLS.

IN BACTERIA THERE ARE SEPARATE, SMALLER DNA MOLECULES WHICH REPLICATE INDEPENDENTLY.

PROCESS BY WHICH BACTERIA TAKE UP BROKEN PIECES OF DNA.

CHAPTER 10

PATTERNS OF INHERITANCE

BEFORE READING ACTIVITIES

I. Before you begin to read, review the important concepts from Chapter 9. Remember that Chapter 9 and Chapter 10 are part of a group which discusses the continuity of life (or cell division and genetics). What are chromosomes? What is the cell cycle? What are the phases of the cell cycle? What is meiosis? Mitosis? What are the components of gamete formation?

II. Getting an overview

1. Turn to page 176. Read the chapter outline about patterns of inheritance.

2. How many major ideas are there in this chapter?

3. How many minor ideas? Which major ideas have minor ideas which you need to learn about?

   1.
   2.

4. Which is a broader concept, "How genes behave" or "Genotype and phenotype"?

5. Construct a graphic organizer to show the relationship between major and minor ideas. Use another sheet of paper and turn it sideways so you can leave more room between ideas.

PATTERNS OF INHERITANCE

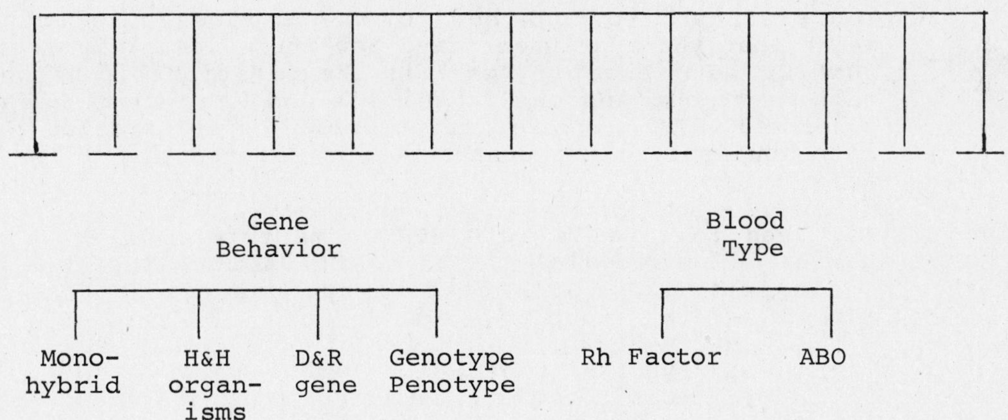

Your organizer should like the one above.
Fill in the headings missing for each line.
Use a different color of ink to write in the lower level headings.
Color coding your graphic organizer will help you keep a better sense of what topic headings work together as a group.

III. Read the short introductory paragraph on page 177. As you do, think about how much the information in this chapter relates to YOU as a separate, unique human being. Read the summary on pages 197 and 198. Separate and read through the vocabulary cards. Finally, take the Post-Test on page 198.

IV. Read the Learning Objectives on page 176.

1. As you read these objectives, underline the word that tells you what you need to do: define, relate, summarize, etc.

2. Review what these words mean; use your memory or turn back to the earlier chapters in this text.

3. Go through the text and find out which pages will give you the information you need to accomplish each objective. Write the page numbers next to the objective.

V. Vocabulary

1. Sort the terms into groups that fit together; that is, categorize them.

2. Determine which terms you already know parts of. For example, many of the terms in this chapter use prefixes that you probably know the meanings of: non-, trans-, co-, in-, hypo-, homo-, hetero-, auto-, hemi-. If you don't know these prefixes, learn them so you will be able to use them to define any new word which incorporates them.

DURING READING ACTIVITIES

I. Using direction-giving words

1. Remember that you need to be able to summarize and describe ideas.

2. You also need to know how to RELATE ideas. Relating ideas means that you must understand how these ideas fit together or CONNECT. For example, what is the relationship between what happens to chromosomes in meiosis and how genetic traits are inherited? As you read, try to find the information that will help you explain the connection or relationship between facts and concepts.

3. You need to be able to DEFINE some terms. When you define terms you should give the MAJOR CHARACTERISTICS of the term. It is important to be concise but you need to include ALL important ideas about the term.

II. Continue to mark your textbook.

1. Differentiate between main ideas and minor ideas.
2. Circle new vocabulary and underline the definition.
3. Use asterisks to indicate ideas of special significance.
4. If you do not understand something, put a ? in the margin and then ASK THE INSTRUCTOR what it means.

AFTER READING ACTIVITIES

I. Review your textbook marking. Make any needed corrections so you will not need to reread the text when you study for an examination.

II. Test yourself on the new vocabulary; that is, see if it is already stored in your long-term memory.

1. Check to see if your categorization of terms (grouping) was appropriate. If you put some terms into groups that were not appropriate, change them.
2. Have you related the new vocabulary to things you already know? For example, what do "dominant" and "recessive" have to do with what you already know about the continuity of life or dominant people?

3. Practice reciting the terms and their definitions. Either say the information aloud or to yourself, over and over again. Remember to practice these terms in groups of 5-7 at any one time.

III. Go back to the graphic organizer you made before you began reading this chapter. Explain to yourself or another person what concepts are involved in each of the headings and subheadings. Fill in as many details as you can and think about how each of these parts of the inheritance pattern relates to you and each other human being.

IV. Answer the following questions.

1. What is Mendel's idea about how heredity is passed from one generation to the next?

2. What are the three laws of inheritance developed by Mendel?

3. The law of ____________ ____________ is concerned with the influence of one heritable trait on the inheritance of another trait.

4. The law of ____________ is concerned to the distribution of individual genes from their corresponding gene pairs which occurs during meiosis.

5. In most individuals two alternative forms of each gene are present, yet only one of these alternatives is expressed. This is the law of ____________.

6. Examine Figure 10-2. From this diagram describe what is meant by a gene locus.

7. Examine Figure 10-3. What important biological process accounts for the separation of the alleles for coat color so that each gamete has only one gene for coat color?

8. Study the example in Figure 10-4, then attempt to work the following problem. In mice long fur is dominant over short fur. You want to cross two mice, each is heterozygous for long fur. What percentage of the offspring will have short fur? Let F = long fur and f = short fur.

9. How would you name the generation of offspring produced by the F_2 generation?

10. Are all genes either clearly dominant or clearly recessive?

11. Examine Figure 10-5. What phenomenon is depicted in this diagram?

12. You have decided to raise guinea pigs and start your guinea pig family with a brown female and a white male. For each of the situations below, indicate which gene (characteristic) is dominant over its allele, or whether there is incomplete dominance.

    A. If all the offspring are white.

    B. If all the offspring are tan.

    C. If all the offspring are brown.

13. Examine Figure 10-6 which illustrates independent assortment. Two traits are being studied.

    A. What are all the possible genotypes of the sperm?

    B. What are all the possible genotypes of the eggs?

    C. What is the most frequently observed phenotype?

    D. What is the least frequently observed phenotype?

14. Are any human traits inherited in a polygenic fashion? If yes, name one.

15. Examine Figure 10-7 and provide the missing information. According to the author's examplantion, who would have darker skin color, an individual with a genotype AaBBcc or an individual with a genotype AAbbCC.

16. With respect to the sex chromosome, what is the genotype of a human male? Of a human female?

17. What kind of genes are contained on the very small Y chromosome?

18. Does the X chromosome carry genes which confer female characteristics to the individual? Explain.

19. Examine Figure 10-8. On theoretical grounds, what percentage of a large human population should be female?

20. Examine Figure 10-9. In a mating of a color blind father and a mother who is heterozygous for the colorblindness gene, what percent of the male offspring will be colorblind?

21. Examine Figure 10-11. Is it possible for one of the offspring of this mating to be a secretor and have normal muscle function?

22. If a woman is Rh negative, can she have other children without having to worry about erythroblastosis fetalis?

23. Examine Table 10-1, the ABO blood types, and answer the following questions.

    A. If the phenotype is type A blood, what are the possible genotypes?

    B. If the blood is type A, what antibodies are found, and where are they found?

C. If the blood type is O, what antigens are present, and where are they found?

24. The three alleles which determine blood type are _____, _____, and _____.

25. Examine the mating shown in Figure 10-14; carry out a similar cross for an AB individual and an O individual, and list all the possible phenotypes.

26. Examine Table 10-2. List three disorders that are autosomal recessive.

27. Examine Table 10-3 showing some chromosome abnormalities and answer the following questions.

A. What is the usual result of a trisomy condition?

B. What accounts for the production of these trisomy conditions?

C. What is the consequence of a trisomy X condition?

D. What is the probability of having a Down's Syndrome child?

28. Each person carries approximately ____________ harmful, recessive genes.

Chapter 10 - Answers

1. Heredity is passed by factors (genes) which exist in pairs.
2. Law of segregation, law of dominance, law of independent assortment.
3. Independent assortment
4. Segregation
5. Dominance
6. The definite position on a chromosome occupied by a gene
7. Meiosis
8. 75% long fur, 25% short fur
9. F_3
10. No
11. Incomplete dominance
12. A. White is dominant over brown
    B. Incomplete dominance
    C. Brown is dominant
13. A. EC, Ec, eC, ec
    B. EC, Ec, eC, ec
    C. Free, curling
    D. Attached, noncurling
14. Yes. Skin color
15. AAbbCC
16. XY, XX
17. Genes conferring male characteristics
18. No, both males and females have X chromosomes
19. 50%
20. 50%
21. Yes. ss Mm
22. Yes. Now Rh negative mothers are treated after birth of first child to prevent any problems.
23. A. AA, AO
    B. Anti-B, found in the blood plasma
    C. No antigens, but if they existed, they would be on the red blood cell membrane.
24. A, B and O
25. Possible phenotypes are Type A and Type B
26. See Table 10-2
27. A. Many trisomy conditions lead to early death, although many trisomy 21 individuals live well into their 30's.
    B. Abnormal meiosis
    C. Does not seem to cause too many problems
    D. Generally 1/700, although frequency increases with the age of the mother.
28. Eight

MEDNEL'S LAWS OF INHERITANCE

MENDEL'S IDEA OF HEREDITY

LAW OF SEGREGATION

LAW OF DOMINANCE

LAW OF INDEPENDENT ASSORTMENT

LOCUS

ALLELES

DOMINANT GENE

RECESSIVE GENE

MONOHYBRID CROSS

PUNNETT SQUARE

PARENTAL GENERATION, P1

HEREDITY IS TRANSMITTED BY UNIT FACTORS, WHICH EXIST IN PAIRS.

CONCLUSIONS MENDEL REACHED CONCERNING THE MECHANISMS OF HEREDITY.

WHEN TWO ALTERNATIVE FORMS OF THE SAME GENE ARE PRESENT IN AN INDIVIDUAL, ONLY ONE OF THE ALTERNATIVE IS USUALLY EXPRESSED.

WHEN GAMETES (EGGS AND SPERM) ARE FORMED, THE TWO GENES OF EACH GENE PAIR SEPARATE FROM ONE ANOTHER. EACH GAMETE RECEIVES ONLY ONE GENE OF EACH PAIR.

THE GENE FOR A PARTICULAR TRAIT OR CHARACTERISTIC OCCURS AT A PARTICULAR POSITION IN THE CHROMOSOME.

IF ONE CONSIDERS TWO OR MORE INDEPENDENT CHARACTERISTICS IN A GENETIC CROSS, EACH CHARACTERISTIC IS INHERITED WITHOUT RELATION TO OTHER TRAITS. ALL POSSIBLE COMBINATIONS OF TRAITS CAN OCCUR. TRUE ONLY IF TRAITS ARE CARRIED ON NONHOMOLOGOUS CHROMOSOMES.

THE GENE OF AN ALLELIC PAIR WHICH EXPRESSES ITSELF.

TWO OR MORE ALTERNATIVE FORMS OF THE GENES GOVERNING THE SAME TRAIT OCCUPY CORRESPONDING LOCI ON HOMOLOGOUS CHROMOSOMES.

A CROSS BETWEEN TWO INDIVIDUALS THAT DIFFER WITH RESPECT TO A SINGLE CHARACTERISTIC.

THE GENE OF AN ALLELIC PAIR THAT DOES NOT EXPRESS ITSELF IN THE PRESENCE OF A DOMINANT ALLELE.

GENERATION A PARTICULAR GENETIC EXPERIMENT IS BEGUN.

MEANS OF REPRESENTING THE PROBABLE COMBINATIONS OF EGGS AND SPERM. LETTERS ARE USED TO REPRESENT PARTICULAR TRAITS, AND THE GENOTYPE OF EACH POSSIBLE ZYGOTE IS INDICATED IN EACH SQUARE.

F_1

F_2

DOMINANT

RECESSIVE

HOMOZYGOUS

HETEROZYGOUS

GENOTYPE

PHENOTYPE

INCOMPLETE DOMINANCE

DIHYBRID CROSS

POLYGENES

AUTOSOMES

SECOND FILIAL GENERATION, OFF-SPRING OF THE F_1 GENERATION.

FIRST FILIAL GENERATION, OFF-SPRING OF THE PARENTAL GENERATION.

WHEN GENE IS EXPRESSED ONLY WHEN IT IS PRESENT IN THE HOMOZYGOUS CONDITION.

WHEN ONE GENETIC OPTION (GENE) AS CONTAINED IN AN ALLELIC PAIR IS ALWAYS EXPRESSED WHEN IT IS PRESENT IN THE HETEROZYGOUS CONDITION.

WHEN EACH OF THE ALLELES OF AN ALLELIC PAIR IS DIFFERENT.

WHEN BOTH ALLELES FOR A PARTICULAR TRAIT ARE THE SAME.

THE APPEARANCE OF THE INDIVIDUAL WITH RESPECT TO A CERTAIN INHERITED TRAIT.

AN INDIVIDUAL'S GENETIC MAKEUP. GENOTYPE IS NOT DIRECTLY DETECTABLE.

A MATING THAT INVOLVES INDIVIDUALS DIFFERING IN TWO TRAITS.

WHEN THE HETEROZYGOTE HAS A PHENOTYPE THAT IS INTERMEDIATE BETWEEN THOSE OF ITS TWO PARENTS.

CHROMOSOMES WHICH EXIST IN PAIRS, AND EACH MEMBER OF THE PAIR IS SIMILAR IN SIZE, SHAPE AND GENE CONTENT TO ITS PARTNER CHROMOSOMES.

SOME TRAITS ARE GOVERNED IN THEIR EXPRESSION BY MORE THAN ONE PAIR OF ALLELIC GENES.

SEX CHROMOSOMES

HEMIZYGOUS

SEX LINKAGE

KARYOTYPE

LINKAGE

AUTOSOMAL LINKAGE

BLOOD TYPE

AGGLUTINATION

HEMOLYSIS

RH POSITIVE BLOOD

PH NEGATIVE BLOOD

MULTIPLE ALLELES

THE Y CHROMOSOME PAIRS WITH A NONHOMOLOGOUS X CHROMOSOME IN HUMAN MALES. GENES ON THE X CHROMOSOME HAVE NO ALLELES ON THE Y CHROMOSOME, AND IN THE MALE THE X CHROMOSOME GENES ARE EXPRESSED. THESE GENES ARE NEITHER HOMOZYGOUS NOR HETEROZYGOUS BUT ARE HEMIZYGOUS.

CHROMOSOMES WHICH CAN EXIST AS A PAIR OF DISSIMILAR CHROMOSOMES. IN HUMANS, FEMALES HAVE A SIMILAR PAIR AND MALES HAVE A DISSIMILAR PAIR. MALES ARE XY AND FEMALES ARE XX.

THE PHOTOGRAPH SHOWING THE PAIRS (USUALLY) OF CHROMOSOMES ARRANGED AS HOMOLOGOUS PAIRS.

REFERS TO CHARACTERISTICS WHICH ARE CARRIED ON THE SEX CHROMOSOMES, THE X OR Y CHROMOSOMES.

GENES WHICH OCCUR TOGETHER ON A NONSEX CHROMOSOME (AUTOSOME), AND WHICH ARE INHERITED TOGETHER.

SITUATION WHERE MULTIPLE GENES ARE PRESENT ON A SINGLE CHROMOSOME. LINKED GENES TEND TO BE INHERITED TOGETHER.

ABNORMAL CLUMPING OF RED BLOOD CELLS WHEN THE RED BLOOD CELL SURFACE PROTEIN ANTIGEN COMBINES WITH A COMPLEMENTARY PROTEIN CALLED AN ANTIBODY.

BLOOD TYPE IS DETERMINED BY THE TYPE OR TYPES OF PROTEINS (ANTIGENS) PRESENT ON THE SURFACE OF THE RED BLOOD CELL MEMBRANE.

WHEN Rh ANTIGENS (PROTEINS) ARE ASSOCIATED WITH THE RED BLOOD CELL MEMBRANE, AND THERE ARE NO ANTI-Rh ANTIBODIES IN THE PLASMA.

COMPLETE BREAKDOWN OF RED BLOOD CELLS.

WHEN A PARTICULAR GENE, WITH A PARTICULAR LOCUS, EXISTS IN MORE THAN TWO FORMS. A NORMAL PERSON POSSESSES NO MORE THAN TWO ALLELES.

WHEN THERE ARE NO Rh ANTIGENS AND NO NATURAL ANTI-Rh ANTIBODIES.

RH INCOMPATABILITY

ERYTHROBLASTOSIS FETALIS

LANDSTEINER'S LAW

INBORN ERRORS OF METABOLISM

ANEUPLOIDY

NONDISJUNCTION

TRANSLOCATIONS

LYON HYPOTHESIS

BARR BODIES

CONSANGUINITY

CONDITION WHEN Rh INCOMPATA-BILITY EXISTS, AND MATERNAL ANTI-Rh ANTIBODIES ENTER FETAL CIRCULATION, CAUSES CLUMPING OF FETAL BLOOD, AND SOMETIMES DEATH OF THE FETUS BEFORE BIRTH.

MOTHER IS Rh NEGATIVE, FATHER IS Rh POSITIVE, AND THE FETUS IS Rh POSITIVE. SOME FETAL BLOOD ENTERS MOTHERS BLOOD, ANTI-Rh ANTIBODIES PRODUCED WHICH CAUSE PROBLEMS IN SUB-SEQUENT PREGNANCIES.

HUMAN DISORDERS INVOLVING ENZYME DEFECTS LINKED WITH GENETIC MUTATIONS.

THE ANTIBODIES OF THE PLASMA ARE NEVER COMPLEMENTARY TO THE ANTIGENS OF THE CELL. POSSIBLE PROBLEM IN BLOOD TRANSFUSIONS.

A DEFECT IN MEIOSIS IN WHICH THE CHROMOSOMES FAIL TO SEPARATE.

CHROMOSOME ABNORMALITY USUALLY INVOLVING AN ABNORMAL NUMBER OF CHROMOSOMES. SOMETIMES PER-SON WILL HAVE THREE CHROMOSOMES PER SET INSTEAD OF THE USUAL TWO (TRISOMY).

DEVELOPED BY GENETICIST MARY LYON. IN NORMAL FEMALES ONE OF THE X CHROMOSOMES IN EACH CELL IS NOT FUNCTIONAL OR NOT FULLY FUNCTIONAL. RANDOM CHOICE AS TO WHICH X CHROMOSOME IS INACTIVATED.

WHEN A PART OF ONE CHROMOSOME BREAKS OFF AND ATTACHES TO ANOTHER CHROMOSOME. RESULTS IN EXTRA GENETIC MATERIAL BE PRESENT IN A CELL.

MARRIAGE BETWEEN RELATIVES. INCREASED RISK OF BEARING A CHILD WITH SOME MAJOR GENETIC PROBLEM.

THE INACTIVATED X CHROMOSOME SEEN IN SOME CELLS.

CHAPTER 11

DNA: THE MOLECULAR BASIS OF INHERITANCE

BEFORE READING ACTIVITIES

I. Before you begin to read, review the important concepts from Chapter 10. Use your graphic organizer to give yourself a self-test. Can you take each of the headings and explain the important concepts. For example, what are Mendel's laws? How do genes behave? Try to put important details into a lower level portion of your organizer; use a different color of ink when you add details for monohybrid level concepts.

II. Getting an overview

1. Turn to page 200. Read the chapter outline about DNA.

2. What is DNA? You should be able to give a general answer from the title of the chapter.

3. How many major ideas are there in Chapter 10? Which major ideas have subsections (minor ideas) you need to learn about?

4. What is the function of the two sections labelled "Focus On. . ."?

5. On a separate sheet of paper, construct a graphic organizer to show the relationship between major and minor ideas. Turn the sheet sideways to leave more room between ideas.

MOLECULAR BASIS OF INHERITANCE = DNA

- Secret Formula
  - Miescher & Nucleus
  - X-ray Eye
  - Building Models
- Structure of DNA
- DNA Replication

Your organizer should look like the one above. Be sure to use a different color of ink for each level of headings so you will see how groups of ideas fit together as a group.

III. Read the vocabulary cards. Read the introductory paragraph on page 201 and the summary on page 212. Use this new information to take the Post-Test on page 212.

IV. Read the Learning Objectives on page 200.

1. Underline the direction word in each objective. The words you should underline are: outline, describe, predict, summarize. Think about what these mean so you will have an idea of what to read for.

2. Go through the text to determine which pages will contain the information you need to accomplish each objective. Write the page numbers next to the objection.

V. Vocabulary

Tear out the vocabulary cards and read through them one more time before you begin reading the chapter carefully. (You should have read them earlier, before you took the Post-Test.) Note that there are only a few new vocabulary terms. Be sure you continue to review the terms from Chapters 9 and 10.

DURING READING ACTIVITIES

I. Using Direction Words

1. Remember that you need to be able to summarize and describe concepts. A new term used in this set of objectives is OUTLINE. When you outline you do use the same basic form as when you summarize and describe; that is, you state the important details of the topic (in this case the history of the scientific investigation of the nucleic acids). When you OUTLINE, you may use a NUMBERED LIST.

2. The other new direction word used is PREDICT. When you see the direction word predict, this should alert you to look for some specific principles BECAUSE predicting means being able to guess what will happen in a new situation using knowledge you already have. In this case, turn to page 208 in your textbook. The last paragraph explains how you can PREDICT a base sequence.

II. Remember to read the figures in the text by first trying to understand the pictorial portion and then reading the verbal portion.

AFTER READING ACTIVITIES

I. Review your textbook marking.

II. Test yourself on the new vocabulary to see if it is stored in your long-term memory. Review the vocabulary from Chapters 9 and 10 at the same time because these chapters are all part of the same unit about the continuity of life.

III. Go back to the graphic organizer you made before you began reading this chapter. Explain to yourself (or to someone else) what concepts are involved in each of the headings. Add in the important details at each level using KEY WORDS only so you will be able to review this information at a later time. Remember to color code by level of importance.

IV. Answer the following questions.

Questions

1. Read the Focus on DNA and the transfer of Genetic Information. What conclusions can be drawn from these experiments?

2. What important technique was used to work out the structure of DNA?

3. Genetic information is specified by the ____________ of the nucleotides in the DNA molecule.

4. Examine Figure 11-2. Do you notice a consistent relationship between the nitrogenous bases on one strand and those on the companion (complementary) strand?

5. Given the following nucleotide sequence of one of the two strands of a DNA molecule, what is the sequence of the partner strand of TACCATTATTCC?

6. Examine Figure 11-5. What is the relationship between the bases in an old strand and those in a new strand?

7. The mode of DNA duplication in organisms is ______________ and ______________.

8. The enzymes involved in DNA duplication are the ______________ ______________. These enzymes need a ____________ and the four ______________ which are assembled into DNA.

Chapter 11 - Answers

1. That DNA is the genetic material that can cause transformation.
2. X-ray diffraction analysis.
3. Sequence
4. A always bonds to T, G always bonds to C
5. ATGGTAATACGG
6. They are complementary
7. Semiconservative, bidirectional
8. DNA polymerases, template, nucleotides

DOUBLE HELIX

NUCLEOTIDES

BASE PAIRING

ANTIPARALLEL

DNA REPLICATION

SEMICONSERVATIVE REPLICATION

BIDIRECTIONAL DNA REPLICATION

DNA POLYMERASES

THE SUBUNITS WHICH COMPRISE DNA. EACH NUCLEOTIDE CONSISTS OF A PENTOSE SUGAR MOLECULE, A PHOSPHATE MOLECULE, AND A NITROGENOUS BASE--EITHER CYTOSINE, THYMINE, ADENINE OR GUANINE.

STRUCTURE OF DNA AS PROPOSED BY WATSON AND CRICK. DNA CONSISTS OF TWO INTERTWINED MOLECULAR HELICES.

THE TWO STRANDS IN A DOUBLE STRANDED DNA MOLECULE RUN IN OPPOSITE DIRECTIONS.

THE PAIRING ARRANGEMENT IN DNA MOLECULES SUCH THAT A GUANINE ON ONE STRAND IS BONDED TO A CYTOSINE ON THE COMPLEMENTARY STRAND. SIMILARLY, ADENINE IS BONDED TO THYMINE.

DNA CHAINS SEPARATE, AND EACH ONE SERVES AS A TEMPLATE FOR THE PRODUCTION OF A *NEW* CHAIN OF COMPLEMENTARY NUCLEOTIDES.

PROCESS BY WHICH DNA MAKES AN EXACT REPLICA BY ITSELF.

THE ENZYMES WHICH SYNTHESIZE DNA UTILIZING AN OLD STRAND AS A TEMPLATE.

DNA REPLICATION (SYNTHESIS) PROCEEDS IN BOTH DIRECTIONS FROM AN INITIATION SITE.

CHAPTER 12

GENE FUNCTION AND REGULATION

BEFORE READING ACTIVITIES

I. Before you begin to read, review the important concepts from Chapter 11: DNA, The Molecular Basis of Inheritance. Use your graphic organizer to test yourself. Take each of the headings and explain the main ideas involved. Be sure to add in the important details which should be in the lower level of the graphic organizer. In addition, go back to the graphic organizers for Chapters 9 and 10 and review the major ideas of each level. Remember that these chapters all make up one part of the textbook which discusses cell division and genetics.

II. Getting an overview

1. Turn to page 214. Read through the outline which gives an overview of the information about gene function and regulation.

2. Note that there are some terms in the outline which you know from other contexts: genetic, code, translation, transcription, control. Try to decide what these words mean in the context of biological science. For example:

   a. What do you think the genetic code is? What is a code of any other type? What are genes? How do these two words probably fit together?

   b. What is "translation" in the vocabulary you already have? Given what you already know about genetics and cell division, what do you think translation will mean in this context?

3. On a separate sheet of paper, construct a graphic organizer to show the relationship between the major and minor ideas. Remember to turn the sheet sideways to leave more room between ideas. Be sure to color code the levels of information so you can easily visualize the relationships.

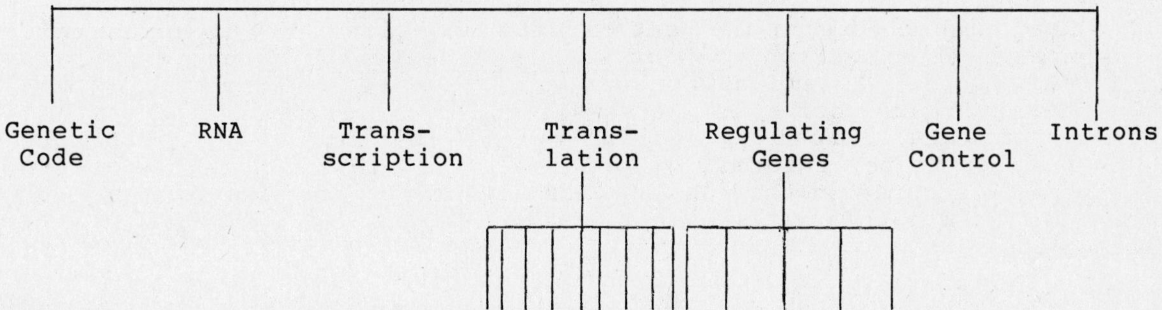

The basic structure of your graphic organizer should look like the one above. A look at this organizer should show you that there are two areas that are most important: translation and regulating genes. To help you picture these concepts with all of their contributing parts, you should make a separate graphic organizer for each of these. Just make the top level structure now and fill in the details as you read the sections.

III. Read through the vocabulary cards. Read through the Introduction and Summary (on pages 233 and 234). Take the Post-Test on page 235.

IV. Vocabulary

Tear out the vocabulary cards and begin to sort them into piles that seem to fit together. Do they seem to be related to the same concept? Do they have word parts that are the same (such as trans-)? Which words do you know from another context: transloca-tion? initiation? regulator?

DURING READING ACTIVITIES

I. Refine your graphic organizer.

1. Remember to read through whole sections and then stop to review. You should read the section about the genetic code on pages 215 and 216, making notes for each paragraph, and then review the important information about the genetic code. As you review, put the key concept terms in your graphic organizer.

2. When you get to the section on translation on page 217, you will note that there are 8 subsections. This is where you should have a separate graphic organizer to help you visualize the concepts involved in translation. Constructing your graphic organizer will help you form a picture of information that needs to be used to understand this one important concept.

3. You should also construct a graphic organizer for the information about regulating genes which begins on page 225.

II. Note that on page 221 there is an "Overview of Translation" and then a Figure on page 222 to help you understand this important process.

III. Read each figure in the text carefully. First, try to understand the graphic portion <u>without reading the text</u>. Second, read the verbal portion and compare it with your verbal interpretation. Third, close your eyes and see if you can visualize each part of the figure. Finally, decide how the information in the figure relates to the information printed in the text. Note that there are 12 figures in this one chapter, so the figures are an important part of the information in the chapter.

AFTER READING ACTIVITIES

I. Review the graphic organizer you made before you began reading this chapter. Explain to yourself what each heading means. Be sure to include important subheadings and color code them so you will recognize the level of importance.

II. Review the graphic organizers you made for the sections on translation and regulating genes. If you cannot explain any topic in these headings, go back to the text and review.

III. Test yourself on the vocabulary. There are a lot of new terms so you will need to spend some time reviewing them in the next few days. Remember that you should study these terms in groups of 5-7 terms because the immediate memory span of an adult is seven "bits" of information so don't overload your brain at any one time.

IV. Answer the following questions.

Questions

1. Why are cellular proteins always needed?

2. How long does it take a cell to make an average-sized protein?

3. In eukaryotic cells, proteins are manufactured in the ____________ of the cell.

4. The three main types of RNA are ______________________________, __________________________, and _________________________.

5. Distinguish between translation and transcription.

6. List the properties of the transcription enzyme RNA polymerase.

7. The enzyme responsible for the transcription is ______________. It produces an ___________ ______________ with a base sequence complementary to that of the DNA from which it was transcribed.

8. During transcription the RNA polymerase molecule attaches to the _____________ _____________ of the DNA molecule. The _____________ _____________ is not ever transcribed in most cases. This means 50 percent of the DNA in a cell is not transcribed.

9. _________________ is the process in which the information contained within an mRNA molecule is utilized to make a particular protein. On the other hand, ___________ is the process in which the genetic information is transferred from a DNA template to an RNA molecule.

10. Transfer RNA molecules carry ____________ ____________ to the ribosomes--the site of protein synthesis.

11. If the mRNA codon is AUG, what is the nucleotide sequence of the anticodon found on the tRNA molecule?

12. Examine Figure 12-3.

   A. What is RNA polymerase?

   B. Aside from forming the RNA molecule, what else does the RNA polymerase do?

   C. How many strands of a DNA molecule are transcribed in a particular region?

   D. Given the following nucleotide sequence of a DNA molecules, what is the nucleotide sequence of the RNA transcribed from this DNA? DNA sequence is TATATTATACGAGCCTTG.

   E. What is the transcribed DNA strand called?

13. Examine Figure 12-4 which shows the structure of tRNA, how many loops are there in a tRNA molecule?

14. Cells have twenty different amino acids, at least twenty different tRNA molecules, and twenty different ____________ ____________ ____________ ____________ molecules. This assures the appropriate amino acid will be attached to its specific tRNA.

15. A ________________ is a cluster of ribosomes bound to an __________ molecule which is actively engaged in _____________ synthesis.

16. Examine Figure 12-6 showing translation.

   A. How many tRNA molecules can be bound to the ribosome at any time?

   B. What are the regions which bind tRNA called?

   C. After the peptide bond forms, what happens to the tRNA in the P-site?

   D. After translocation, what occurs at the A-site?

17. IF-2 and IF-3 are ____________ molecules involved in the ____________ phase of protein synthesis.

18. An ________________ complex consists of EF-Tu, an amino acid - tRNA, GTP all bound to a single ribosome.

19. What happens to the EFTu after the formation of a peptide bond?

20. Examine Figure 12-7. What is the apparent role of IF-3?

21. How many cellular fuel molecules are used to form a single peptide bond?

22. _____________ genes, control _____________, and a set of _________ genes collectively constitute an operon.

23. Examine Figure 12-10 showing the operon model.

    A. What happens when repressor protein binds to the operator site?

    B. What happens when inducer inactivates the repressor protein?

    C. When all the substrate-inducer has been metabolized, what happens?

Chapter 12 - Answers

1. The changing needs of cells to cope with changing environment. To replace damaged proteins.
2. 20 minutes
3. Cytoplasm
4. Messenger RNA (mRNA, ribosomal RNA (rRNA), and transfer RNA (tRNA).
5. Transcription is the production of RNA off of DNA molecules which dictate the sequence of nueleotides in the RNA molecule. Translation is the synthesis of proteins on ribosomes.
6. Recognizes cues in the DNA molecule: (1) recognizes which of the two DNA strands to transcribe; (2) where on the DNA strand to begin transcription, and (3) where it should stop transcription.
7. RNA Polymerase, RNA molecule
8. Coding strand, noncoding strand
9. Translation, transcription
10. Amino Acids
11. UAC
12. A. Enzyme which synthesizes RNA
    B. Unwinds the double stranded DNA molecule.
    C. One
    D. AUAUAAUAUGCUCGGAAC
    E. Template
13. Three
14. Amino acyl tRNA snythetase
15. Polyribosome, mRNA, polypeptide (protein)
16. A. Two
    B. A-site and P-site
    C. It moves away
    D. A new amino acyl-tRNA moves in.
17. Protein, initiation
18. Initiation
19. Nothing, these factors are reutilized.
20. IF-3 maintains the ribosomal subunits. It dissociates from the small subunit and an initiation complex forms.
21. Three
22. Regulatory, elements, structural
23. A. The structural genes are not transcribe
    B. The enzymes A, B, and C are produced.
    C. The repressor protein again binds to the operator.

GENES

TRIPLETS

RIBONUCLEIC ACID (RNA)

MESSENGER RNA (MRNA)

TRANSCRIPTION

RIBOSOMES

RIBOSOMAL RNA

TRANSFER RNA

TRANSLATION

RNA POLYMERASE

PROMOTOR

CODING STRAND (PLUS STRAND)

A COMBINATION OF THREE NUCLEOTIDES WHICH CONTAIN THE INFORMATION TO SPECIFY ONE PARTICULAR AMINO ACID, SINCE THERE ARE FOUR DIFFERENT NUCLEOTIDES, THERE ARE 64 DIFFERENT COMBINATIONS OF THREE NUCLEOTIDES - 64 DIFFERENT TRIPLETS.

CODE FOR ALL THE PROTEINS IN THE CELL.

THIS MOLECULE IS A "MOLECULAR XEROX" COPY OF THE GENETIC INFORMATION CONTAINED IN A SEGMENT OF DNA. THIS GENETIC INFORMATION IS CARRIED TO THE CYTOPLASM WHERE IT ASSOCIATES WITH THE RIBOSOMOS.

THE OTHER TYPE OF NUCLEIC AND BESIDES DNA, DIFFERS FROM DNA. IT CONTAINS THE SUGAR RIBOSE INSTEAD OF DEOXYRIBOSE. COMPOSED OF SEQUENCES OF FOUR DIFFERENT BASES. IT HAS THE BASE URACIL IN PLACE OF THYMINE WHICH HYDROGEN BONDS TO ADENINE.

PARTICLES COMPOSED OF RIBOSOMAL RNA AND PROTEINS (IN APPROXIMATELY EQUAL AMOUNTS). THESE STRUCTURES ASSEMBLE THE AMINO ACIDS IN A SPECIFIC ORDER ACCORDING TO THE INFORMATION CONTAINED IN THE mRNA MOLECULE.

THE SYNTHESIS OF RNA BY AN ENZYME UTILIZING THE INFORMATION CONTAINED IN THE DNA NUCLEOTIDES. THE INFORMATION IS "SPELLED OUT" BY THE SEQUENCE OF THE NUCLEOTIDES. THE BASE SEQUENCE IN DNA IS COMPLEMENTARY TO THE BASE SEQUENCE IN RNA.

THESE ARE A POPULATION OF APPROXIMATELY 60 DIFFERENT SMALL RNA MOLECULES. THERE IS AT LEAST ONE SPECIFIC tRNA MOLECULE FOR EACH OF THE 20 DIFFERENT AMINO ACIDS. THESE tRNA MOLECULES CARRY THE AMINO ACIDS TO THE RIBOSOMES WHERE PROTEINS ARE SYNTHESIZED.

THE RNA COMPONENT OF RIBOSOMES. SEE VOCABULARY CARD FOR RIBOSOMES.

THE NAME GIVEN TO THE ENZYMES WHICH TRANSCRIBE RNA. THIS ENZYME RECOGNIZES A PARTICULAR START SIGNAL ON THE DNA AND STARTS PRODUCING RNA FROM THAT POINT.

THE PROCESS OF PROTEIN SYNTHESIS ON RIBOSOMES.

ONE OF THE TWO DNA STRANDS. THIS ONE IS TRANSCRIBED INTO RNA. THE BASE SEQUENCE IN RNA IS COMPLEMENTARY TO THE BASE SEQUENCE IN THE DNA CODING STRAND.

A PARTICULAR REGION ON THE DNA MOLECULE WHICH IS RECOGNIZED BY THE RNA POLYMERASE AS A START SIGNAL FOR TRANSCRIPTION. CONSISTS OF PARTICULAR SEQUENCE OF NUCLEOTIDES.

NONCODING STRAND (MINUS STRAND)

TRANSLATION

ANTICODON

AMINO ACYL SYNTHETASES

RIBOSOME

POLYRIBOSOMES

A-SITE

P-SITE

TRANSLOCATION

INITIATION

INITIATION FACTOR 3 (IF-3)

INITIATION COMPLEX

THE PROCESS IN WHICH THE GENETIC INFORMATION WHICH HAS BEEN TRANSCRIBED WITHIN AN mRNA MOLECULE DIRECTS THE SYNTHESIS OF A SPECIFIC POLYPEPTIDE.

THE NONTRANSCRIBED STRAND OF A DOUBLE STRANDED DNA MOLECULE.

A SERIES OF ABOUT 20 DIFFERENT ENZYME MOLECULES EACH OF WHICH ATTACHES A SPECIFIC AMINO ACID TO ITS CORRESPONDING tRNA MOLECULE. EACH ENZYME BINDS AN ATP, AN AMINO ACID, AND A tRNA.

THE THREE NUCLEOTIDES WHICH ARE IN A PARTICULAR REGION OF THE tRNA MOLECULE WHICH INTERACT WITH THREE mRNA NUCLEOTIDES. THE CODON - ANTICODON INTERACTION INVOLVES COMPLEMENTARY BASE PAIRING.

TEMPORARY STRUCTURE FORMED BY THE ATTACHMENT OF 30-40 RIBOSOMES TO AN mRNA MOLECULE FOR PROTEIN SYNTHESIS. EACH RIBOSOME IN THIS CLUSTER OF RIBOSOMES PRODUCES THE COMPLETE POLYPEPTIDE ENCODED IN THE mRNA.

SPECIALIZED CYTOPLASMIC ORGANELLE COMPOSED TO TWO RNA AND PROTEIN SUBUNITS. THE RIBOSOME SERVES AS THE SITE FOR AMINO ACID ASSEMBLY INTO POLYPEPTIDES. BINDS mRNA, READS THE INFORMATION IN THE mRNA AND CATALYZES THE PEPTIDE BOND FORMATION BETWEEN TWO AMINO ACIDS.

A REGION ON A RIBOSOME WHICH RETAINS THE tRNA OF THE PEPTIDE AMINO ACID.

A REGION ON A RIBOSOME WHICH ACCEPTS A NEW tRNA-AMINO ACID COMPLEX.

PROCESS WHICH PRECEDES FORMATION OF THE FIRST PEPTIDE BOND. INVOLVES BINDING OF THE mRNA TO THE RIBOSOME AND FORMATION OF A BOND WITH THE FIRST AMINO ACID-tRNA.

MOVEMENT OF THE RIBOSOME DOWN THE mRNA MOLECULE BY ONE CODON.

COMPLEX FORMED BETWEEN mRNA, A SMALL RIBOSOMAL SUBUNIT, A MOLECULE OF THE PROTEIN INITIATION FACTOR 2, AND A SPECIAL tRNA-AMINO ACID COMPLEX--A METHIONINE AND ITS tRNA. ALL INITIATION COMPLEXES BIND AT AUG SEQUENCE OF mRNA.

A PROTEIN FACTOR WHICH CAUSES DISSOCIATION OF THE RIBOSOME INTO ITS TWO SUBUNITS.

ELONGATION COMPLEX

RELEASE FACTORS

FEEDBACK INHIBITION

ALLOSTERIC SITE

OPERON

STRUCTURAL GENES

REGULATOR GENE

OPERATOR SITE

ENZYME INDUCTION

PROTEINS WHICH RECOGNIZE THE TERMINATION CONDONS UGA, UAG, AND UAA, AND BREAK THE BOND BETWEEN THE POLYPEPTIDE AND THE tRNA TO WHICH IT IS BOUND.

COMPLEX OF THE PROTEIN EF-Tu, GTP AND THE AMINO ACID-tRNA ALL BOUND TO THE RIBOSOME.

A SITE ON AN ENZYME IN ADDITION TO THE ACTIVE SITE WHICH REGULATES THE ACTIVITY OF THE ENZYME.

PROCESS IN WHICH THE FIRST ENZYME IN A MULTISTEP ENZYME CATALYZED MULTIREACTION PATHWAY IS REGULATED BY THE PRODUCT OF THIS PATHWAY. THE PRODUCT BINDS TO AN ENZYME AND TEMPORARILY STOPS ENZYME ACTIVITY.

GENES WHICH CODE FOR PROTEINS NEEDED BY THE CELL, SUCH AS ENZYMES AND STRUCTURAL PROTEINS.

A REGULATED UNIT OF GENE EXPRESSION INCLUDING REGULATORY GENES, CONTROL ELEMENTS (SITES WHERE REGULATOR PROTEINS ACT) AND A SET OF STRUCTURAL GENES.

CONTROL ELEMENT OF OPERON ADJACENT TO STRUCTURAL GENES. IF REPRESSOR BOUND, NO TRANSCRIPTION OF STRUCTURAL GENES.

GENE WHICH CODES FOR REPRESSOR PROTEIN THAT REGULATES THE STRUCTURAL GENES.

WHEN THE PRESENCE OF A PARTICULAR SUBSTANCE CAUSES THE SYNTHESIS OF THE ENZYMES TO INTERACT WITH THAT PARTICULAR SUBSTANCE.

CHAPTER 13

GENETIC FRONTIERS

BEFORE READING ACTIVITIES

I. By now you have probably figured out that you should review the important concepts from the previous chapter(s) before you begin to read the new chapter. This is a critical step in learning because intending to remember is a key factor in learning information. Intending to remember should cause you to review since REVIEW is a critical step in REMEMBERING. A second important reason for reviewing is that whatever you read is understood in light of what you already know. The more you knowabout know about cell division and genetics, the easier it will be for you to understand and learn the new information about genetic frontiers.

II. Getting an overview

1. Getting an overview is an important step because it is one way of helping yourself remember more information for a longer period of time. When you make a graphic organizer, you become familiar with the information and you also begin to understand how each piece of information fits into the overall picture. When you understand how information fits into the picture, you are better able to remember it.

2. Use the outline on page 236 to construct your graphic overview.

GENETIC FRONTIERS

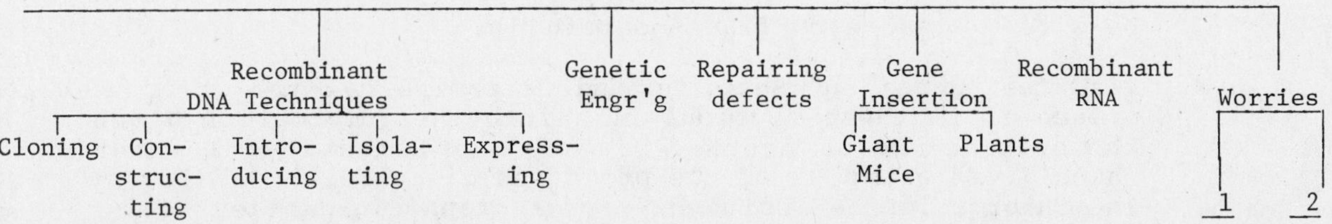

Did you turn your paper sideways to leave more room for the lower level information? Did you color code the levels? Did you shorten the headings so they still make sense but "clue" you in to what information is included in that section? Did you immediately go to the back of the chapter to find out what "Worries" was all about? Can you fill in the two major worries?

3. Read through the vocabulary cards. Note that there are not a lot of new terms. Be sure that you review the terms from Chapters 9, 10, 11 and 12 because all of these terms are likely to be used in this chapter.

4. Read the introduction and summary portions of the text. Take the Post-test on page 253. Did your knowledge of the information from previous chapters help you to fill in the blanks?

DURING READING ACTIVITIES

I. Refine your graphic organizer.

1. You already have an idea of the overview of the whole chapter. Quickly review in your mind the main ideas that will be presented.

2. Carefully read the chapter, section by section. You should read the section on "Recombinant DNA Techniques" paragraph by paragraph and annotate it as you read. Then read the section on "Cleaving DNA" and annotate it as you read. When you get to the end of the section on page 239, review the main ideas presented AND put the major concepts you will need to know on your graphic organizer using a different color. This will clearly help you see that this is information about cleaving which is one of the recombinant DNA techniques which is one of the genetic frontiers. Then you should remember that genetic frontiers are one part of a whole section which discusses the continuity of life. (Go back to the first graphic organizer at the beginning of Chapter 9 if you need to review.)

   Quickly review the information you marked as you read. If you had ? in the margins because you did not understand something, reread the section to see if you understand it better after completing a whole section. IF you DON'T understand, be sure to ask the instructor.

3. Be sure to read each figure carefully.

4. Read the "Focus On" Sections and determine how the information presented in these sections fits into the information given in the body of the chapter. Are they supplementary information which gives examples of the processes discussed? Should any of the information be included in your graphic organizer?

AFTER READING ACTIVITIES

I. Review the graphic organizer and explain the information at each level.

II. Be sure to review the information from all of the chapters in this section of the text.

III. Test yourself on the vocabulary terms from this chapter and from the other chapters dealing with cell division and genetics. Be sure that you practice the vocabulary in groups of 5-7 terms.

IV. Answer the following questions.

Questions

1. What are some of the possible benefits of the emerging recombinant DNA technology?

2. Restriction endonucleuses are ____________ which recognize specific short nucleotide sequences in the ____________ molecule.

3. These enzymes are isolated from ____________ where they function to destroy invading ____________. The ____________ own DNA is protected from destruction by these enzymes.

4. Examine Figure 13-4 showing the insertion of mammalian DNA into a plasmid.

   A. What is accomplished by reverse transcription?

   B. What is the relationship between the nucleotide sequence of the mRNA for insulin and the cDNA mode using this mRNA as a template?

5. Do bacteria have large numbers of plasmids?

6. In order to get expression of the recombinant DNA which is propagated in the bacterial hosts, the inserted gene must have associated ____________ and ____________ genes.

7. Read the Focus on Probing for Genetic Disease. What diagnostic molecules must the investigator have in order to detect the presence of a disease-causing gene?

8. What is one of the important goals of genetic engineering?

9. Read Focus on . . . Molecular Genetics and Cancer.

   A. A cancer cell is one which lacks ____________ ____________ and control.

   B. These cancer cells continue to undergo ____________ or ____________.

10. Is it now possible to introduce a specific gene (piece of DNA) directly into an animal egg and observe expression of the injected gene?

Chapter 13 - Answers

1. Limited only by impact on the society and the environment. These techniques are already allowing the production of human insulin for diabetics and the production of human growth hormone for people lacking this hormone.
2. Enzymes, DNA
3. Bacteria, viruses, bacteria's
4. A. Set the production of a cDNA molecule to the mRNA coding for insulin.
   B. The two nucleotide sequences are complementary.
5. Yes, as many as twenty plasmids per cell.
6. Regulatory, promoter
7. Need both the genomic DNA isolated from any cell to be tested and cDNA probe for the disease in question.
8. One day it may be possible to implant normal, healthy genes into eukaryotic cells to correct defective disease causing genes.
9. A. Biological inhibition
   B. Mitosis or cell division
10. Yes

GENETIC ENGINEERING

RECOMBINANT DNA TECHNIQUE

VECTORING

RESTRICTION ENDONUCLEASES

PLASMID

GENETIC LIBRARY

GENETIC PROBE

ONCOGENES

RETROVIRUS

TECHNIQUES WHICH PERMIT THE FORMATION OF NEW COMBINATIONS OF GENES BY ISOLATING GENES FROM ONE ORGANISM AND INTRODUCING THEM INTO A SIMILAR OR EVEN UNRELATED ORGANISM.

THE MANIPULATION OF DNA OUTSIDE AN ORGANISM SO THAT NEW STRAINS OF ORGANISMS WITH NEW CHARACTERISTICS CAN BE CONSTRUCTED.

ENZYMES WHICH ARE INDIVIDUALLY QUIET SPECIFIC IN BREAKING UP VERY LARGE DNA MOLECULES INTO SMALLER MANAGEABLE PIECES. THESE ENZYMES EACH RECOGNIZE A SPECIFIC SHORT NUCLEOTIDE SEQUENCE WHICH IS THEN CLEAVED.

THE TRANSFER OF GENES FROM ONE ORGANISM TO ANOTHER.

A COLLECTION OF DNA FRAGMENTS GENERATED BY RESTRICTION ENDONUCLEASE TREATMENT OF GENOMIC DNA WHICH ARE REPLICATED ALONG WITH THE HOST BACTERIA CELLS.

SMALL ACCESSORY CHROMOSOMES IN BACTERIA WHICH ARE REPLICATED ALONG WITH THE MAIN CHROMOSOME AND DISTRIBUTED DURING CELL DIVISION. CIRCULAR DNA.

GENES IN CELLS WHICH WHEN EXPRESSED <u>MAY</u> CAUSE CANCER.

A RADIOACTIVELY LABELED SEGMENT OF cDNA (COMPLEMENTARY DNA) WHICH CAN BE USED TO DETECT THE PRESENCE OF A PARTICULAR mRNA MOLECULE.

A CLASS OF RNA VIRUSES WHICH MAKE DNA COPIES OF THEIR OWN RNA DURING THE INFECTION. THESE VIRUSES CARRY OUT REVERSE TRANSCRIPTION.

CHAPTER 14

THE DIVERSITY OF LIFE: MICROBIAL LIFE AND FUNGI

BEFORE READING ACTIVITY

I. You should note that this is the first chapter in a new section of the text with the title "The Diversity of Life." Each chapter in this section discusses some forms of life. Reading through the chapter divisions and then looking at the outline of Chapter 14 on page 257 shows that these chapters must work together to give us a complete picture of the diversity of life. Chapter 14 gives an introduction to the whole section and then discusses three kingdoms. The other two chapters each discuss one kingdom.

II. Read through the chapter outline on page 257 carefully.

1. Note that sections I, II, III, and IV give the overview or general information and that Sections V, VI, VII give more specific information about a particular kingdom.

2. What are the major divisions of the kingdom monera?

3. What are the major divisions of the kingdom protista?

4. How do you think phylum fit into the kingdom protista?

III. Read through the vocabulary cards. Note that there are 60 new terms. Since this is a large number of terms to remember, you need to begin planning how you will learn. The first step should be to organize these terms into groups that make sense to you. For example,

1. All of the terms which tell about the classification system can be put together.

2. All of the terms at each level of the classification system can be grouped together: monera, protista, fungi, plantae, animalia.

IV. Read the Introduction on page 258 and the Summary on pages 282 and 283. Take the Post-Test on page 283.

DURING READING ACTIVITIES

I. Read the chapter in sections. Mark the paragraphs in each section AFTER you read them. Use your own marking system to show what the main ideas are. Be sure to make SMALL, CIRCLED NUMBERS near words which indicate a group of things which are all part of the same idea. For example, the terms in the paragraphs on pages 258 and 259 are all part of the classification system and should be numbered. There are seven terms (see the list under "Category" in Table 14-1. You should also write this list in the margin so you can reinforce your learning and find the list quickly when you review.

II. Begin now to learn these groups of terms which are part of the classification system. One way to help yourself remember is to use MNEMONIC DEVICES; that is, create some sentence or word or phrase that you can associate with the information you need to remember. For example, to learn the classification order of kingdom, phylum, subphylum, class, order, family, genus, and species use these steps:

1. Take the first letter of each term: K,P,S,C,O,F,G,S.

2. Think of a sentence that uses each letter in order such as: King Phillip said, call out for Gus and Sam.

III. Another way to help yourself learn these terms would be to just learn the first letters of each term (K,P,D,C,O,F,G,S) and then use these letters to remember the categories of the classification system.

IV. Review the Table 14-2 about the five kingdoms carefully. Start now to remember these five kingdoms: monera, protista, fungi, plantae, animalia. Can you devise a mnemonic that will help you remember these kingdoms?

V. There are 30 figures in this chapter. Since it take a lot of time to read a figure carefully, begin now to write a summary of the information contained in the figure in the margin so you will NOT need to go back and reread this information a second time.

1. To summarize this information, you should paraphrase the text. For example, Figure 14-22 on page 278 explains what happens when a particular fungus infects a cereal flower and how this can result in causing animals or people to become ill.

AFTER READING ACTIVITIES

I. Reread all of your margin notes and other markings. Be sure that you understand what you marked and that you did not omit important information. Also, if you find some information that is not important, remove your marks in some way.

II. Review the vocabulary. You will need to spend alot of time learning all of this new terminology. As you practice saying the definition of each term to yourself, think about how this term is related to some of the other terms and concepts you need to know. For example, bacteria, cocci, bacilli, spirilla can be learned as a group. Gram-positive and gram-negative would be learned together.

III. Answer the following questions.

Questions

1. Examine Table 14-1. List the seven categories of the hierarchal classification scheme beginning with the most inclusive (broadest) category.

2. What category does Felidae belong to? Felis? Chordata? Primates? Plantae? Mays?

3. Examine Figure 14-3 and Table 14-2. Identify and distinguish among the five kingdoms.

   A.

   B.

   C.

   D.

   E.

4. ____________ are organisms which cannot survive by itself.

5. Examine Table 14-3. At which era, period and epoch did each of the biological events take place?

   The emergance of:

   Algae

   Fish

   Terrestrial plants

   Reptiles

   Insects

   Mammals

   Angiosperms

   Birds

   Humans

   The extinction of:

   Dinosaurs

   Some large mammals

6. The organisms which lack nuclear membranes are called ________. They include cyanobacteria and bacteria and are assigned to the kingdom ____________.

7. Bacteria can be categorized by shape. Spiral shaped ones are called _____________. Those which are rod-shaped are called ____________. ________________ shaped are called cocci.

8. Some bacteria have cell membranes which are called ____________ due to their infolding. Others have hair-like structures called ____________.

9. Identify the four types of harmful Eubacteria.

A.

B.

C.

D.

10. Why should Archaebacteria be placed in a separate kingdom?

11. Distinguish among the protozoans based on their locomotory apparatus and special features.

12. Distinguish between all forms of fungi giving characteristics:

Chapter 14 - Answers

1. Kingdom, plylum, class, order, family, genus, species
2. Family, genus, plylum, order, kingdom, species
3. A. Monera - prokaryotes - bacteria - decomposers
   B. Protista - eukaryotes - protozoa - algae - producers and decomposers
   C. Fungi - multicellular molds, yeasts, mushrooms, decomposers
   D. Plantae - multicellular plants - producer
   E. Anamalia - multicellular animals - consumer
4. Viruses
5. Precambrian
   Paleozoic - Ordovizian
   Paleozoic - Devonian
   Paleozoic - Pennsylvanian
   Paleozoic - Silurian
   Mesozoic - Triassic
   Mesozoic - Cretaceous
   Lenozoic - Tertiary - Eocene
   Lenozoic - Ouarternary - Recent
   Mesozoic - Cretaceous
   Cendzoic - Tertiary - Pleistocene
6. Prokaryotes, Monera
7. Spirilla, bacilli, spherical
8. Mesosomes, pili
9. Spirochaetes - syphillis, mycoplasmas, Richettsia - spotted fever, pathogenic bacteria
10. Archaebacteria are unlike eubacteria and cyanobacteria. The archaebacteria do not have peptigoglycan in their cell walls and are biochemically unlike other bacteria.
11. Sarcodina move by means of pseudopodia, Flagellata move by means of flagella, Ciliata move by means of cilia, Sporozoa have no special structures to move--they glide or use body movements
12. Mycorrhizae - symbiotic with roots
    Lichens - symbiotic with green algae or cyanobacteria
    Water molds - aquatic spheres
    Zygomycetes - terrestrial dry spores
    Sac fungi - spores in a sac
    Club fungi - spores in expanded basidia
    Slime molds - protoplasmic masses

TAXONOMY

BINOMIAL NOMENCLATURE

GENUS (PL. GENERA)

SPECIES (PL SPECIES)

FAMILY

ORDER

CLASS

PHYLA (ANIMALS)
DIVISION (PLANTS)

KINGDOM

FIVE KINGDOM SYSTEM OF CLASSIFICATION

VIRUS

ERA, PERIOD, EPOCH

THE LINNAEAN SYSTEM OF USING A TWO-PART NAME FOR ALL ORGANISMS.

THE PRACTICE OF NAMING AND CLASSIFYING ORGANISMS INTO A HIERARCHY.

THE MOST EXCLUSIVE (NARROWEST) OF THE HIERARRCHAL CATEGORIES AND THE SECOND PART OF THE BINOMIAL.

A GROUP OF LIKE SPECIES AND THE FIRST PART OF THE BINOMIAL.

A GROUP OF LIKE FAMILIES WHICH ARE CLOSELY RELATED.

A GROUP OF LIKE GENERA WHICH ARE CLOSELY RELATED.

A GROUP OF LIKE CLASSES WHICH ARE CLOSELY RELATED.

A GROUP OF LIKE ORDERS WHICH ARE CLOSELY RELATED.

THE SYSTEM OF CLASSIFICATION WHICH RECOGNIZES FIVE SEPARATE KINGDOMS - PLANT, FUNGI, ANIMALS, PROTISTS AND MONERANS.

THE HIGHEST (BROADEST) CATEGORY OF LIKE PHYLA WHICH ARE CLOSELY RELATED.

GEOLOGIC TIME INTERVALS WHICH PROVIDE A TIME FRAME REGARDING WHEN ORGANISMS CAME INTO BEING.

A NON-CELLULAR FORM OF LIFE. CONSISTS OF A NUCLEIC ACID CORE, RNA OR DNA AND A PROTEIN COAT.

PRION

PROKARYOTE

EUKARYOTE

BACTERIA

COCCI (A COCCUS)

BACTILLI (A BACILLUS)

SPIRILLA (A SPIRILLUM)

GRAM - POSITIVE

GRAM NEGATIVE

PEPTIDOGLYCAN

MESOSOMES

PILI

MONERANS INCLUDING CYANOBACTERIA AND BACTERIA WHICH DO NOT POSSESS A NUCLEAR MEMBRANE OR OTHER MEMBRANE BOUND ORGANELLES.

AN ORGANISM SMALLER THAN A VIROID WHICH HAS ONLY PROTEIN AS ITS ORGANIC MOLECULES.

UNICELLULAR, EUKARYOTIC ORGANISM FOUND ALMOST IN EVERY DESCRIBED HABITAT.

PROTISTANS INCLUDING ALGAL PROTISTS AND PROTOZOA WHICH DO HAVE A "TRUE" NUCLEUS CONTAINED WITHIN A NUCLEAR MEMBRANE.

ROD-SHAPED BACTERIA.

SPERICAL-SHAPED BACTERIA.

CELL WALL STRUCTURE STAINS VIOLET IN CERTAIN BACTERIA.

SPIRAL-SHAPED BACTERIA.

AN ORGANIC MOLECULE FOUND IN THE CELL WALL OF BACTERIA.

CELL WALL STRUCTURE DOES NOT STAIN IN CERTAIN BACTERIA.

HAIRLIKE STRUCTURES ON THE SURFACE OF MANY GRAM-NEGATIVE BACTERIA.

INFOLDING OF THE CELL MEMBRANE OF SOME BACTERIA.

SPIROCHETES

MYCOPLASMAS

RICHETTSIAS

CHEMOSYNTHETIC AUTOTROPHS

PATHOGENIC BACTERIA

CYANOBACTERIA

PROTISTA

PROTOZOA

ZOOPLANKTON

SARCODONA

PSEUDOPODIA

FLAGELLATA

SMALL BACTERIA LACKING A TYPICAL CELL WALL, POSSESSES A PLIABLE CELL MEMBRANE.

SPECIAL TYPE OF SPIRAL-SHAPED BACTERIA WITH FLEXIBLE CELL WALLS.

BACTERIA WHICH CAN PRODUCE THEIR OWN FOOD.

SMALL BACTERIA INCAPABLE OF CARRYING ON METABOLISM BY THEMSELVES, THEY ARE PARASITIC.

BACTERIA, MANY OF WHICH ARE BLUE-GREEN. MOST FORM LONG RILILIONS OF CELLS OR GLOBULAR COLONIES.

DISEASE CAUSING BACTERIA.

"FIRST ANIMALS" UNICELLULAR HETEROTROPHIC WHICH ARE PRIMARILY INGESTORS (CONSUMERS).

THE KINGDOM OF UNICELLULAR EUKARYOTIC ORGANISMS INCLUDING ALGAL PROTISTS AND ANIMAL PROTOZOA

PHYLUM OF PROTOZOANS WHICH MOVE BY MEANS OF PSEUDOPODIA, LIKE AMEBA (SEE FIGURE 14-11).

SMALL, MICROSCOPIC ANIMAL-LIKE PROTOZOANS WHICH SERVES AS A RICH SOURCE OF ENERGY IN MANY AQUATIC FOOD CHAINS.

PHYLUM OF PROTOZOAMS WHICH MOVE BY MEANS OF FLAGELLA, LIKE <u>EUGLENA</u> (SEE FIGURE 14-14A).

"FALSE FOOT" LIKE EXTENSION OF THE CYTOPLASM WHICH HELPS PROPEL THE SARCODINES (SEE FIGURE 14-11).

CILIATA

SPOROZOA

ALGAE PROTISTS

PHYTOPLANKTON

DINOFLAGELLATES

DIATOMS

FUNGI

YEAST

MOLD

HYPHAE

COENOCYTIC

SEPTA

PHYLUM OF PROTOZOAN WHICH MOVE BY A VARIETY OF MECHANISMS. POSSESS SPORE LIKE STAGES IN SOME PART OF THEIR LIFE HISTORY, LIKE PLASMODIUM, YELLOW FEVER (SEE FIGURE 14-16).

PHYLUM OF PROTOZOANS WHICH MOVE BY MEANS OF MANY HAIR-LIKE STRUCTURES CALLED CILIA, LIKE PARAMECIUM (SEE FIGURE 14-15b).

SMALL MICROSCOPIC PHOTOSYNTHETIC PROTOZOAN WHICH MAKE UP THE FREE FLOATING PHOTOSYNTHETIC FOOD RESOURCE IN AQUATIC ENVIRONMENTS.

UNICELLULAR AUTOTROPHS.

TYPE OF ALGAL PROTIST WHICH HAS A SILICON CELL WALL WHICH ARE USED COMMERCIALLY IN INSULATING MATERIAL, SWIMMING POOL FILTERS AND TOOTHPASTE.

TYPE OF ALGAL PROTIST WHICH CAUSES RED TIDE.

A UNICELLULAR FUNGI.

A KINGDOM OF ORGANISMS COMPRISED OF MOLDS, YEASTS AND MUSHROOMS. SOME ARE UNICELLULAR, MOST ARE MULTICELLULAR WHICH ABSORB THEIR FOOD.

THREADLIKE STRING OF CELLS OF A MOLD, COLLECTIVELY FROM A TISSUE-LIKE MASS CALLED MYCELIUM.

A MULTICELLULAR FUNGI, INCLUDES MILDEW, RUSTS, SMUTS, SLIME MOLDS AND MUSHROOMS.

NONE-DIVIDED CELL WALLS IN HYPHAE.

CELL WALLS WHICH DIVIDE HYPHAE INTO CELLS.

LICHENS

MYCORRHIZAE

ZYGOMYCETES

WATER MOLD

CLUB FUNGI

SAC FUNGI

BASIDIOSPORES

BASIDIA

SLIME MOLDS

FRUITING BODIES

A TYPE OF FUNGI WHICH BECOME SYMBIOTIC WITH THE ROOTS OF PLANTS.

A TYPE OF SYMBIOTIC RELATIONSHIP BETWEEN FUNGI AND AN ALGAE.

MOLDS WHICH HAVE MOBILE ASEXUAL SPORES, AQUATIC, LIKE POTATOE BLIGHT.

TERRESTRIAL FUNGI PRODUCING RESTING SPORES CALLED ZYGOSPORES, LIKE BREAD MOLDS.

FUNGI WHICH PRODUCE SPORES IN A SAC, LIKE DUTCH ELM DISEASE.

CLASS BASIOIOMYCETES FORMING THEIR HYPHARE INTO LARGE CLUB-LIKE FORMS, LIKE MUSHROOMS.

ENLARGED CLUB-SHAPED HYPHAEL CELLS.

RESIDE AT THE BASE OF BASIDIA.

THE HYPHAE OF THE MYCELLIUM, THE STRUCTURE OF A MUSHROOM.

MOLDS WHICH DEVELOP AS STREAMING MASSES OF HYPHAE FEEDING ON DECAYED ORGANIC MATTER.

CHAPTER 15

PLANT LIFE

BEFORE READING ACTIVITIES

I. Review Chapter 14. Use the outline on page 257 to recite what you know about each topic. This is a good time to start a graphic organizer which will include all five kingdoms. To build this graphic organizer, you should get a larger sheet of paper and remember that you won't have the information to fill in all of the lower level information until you complete Chapter 16. The basic structure of your organizer should look like this:

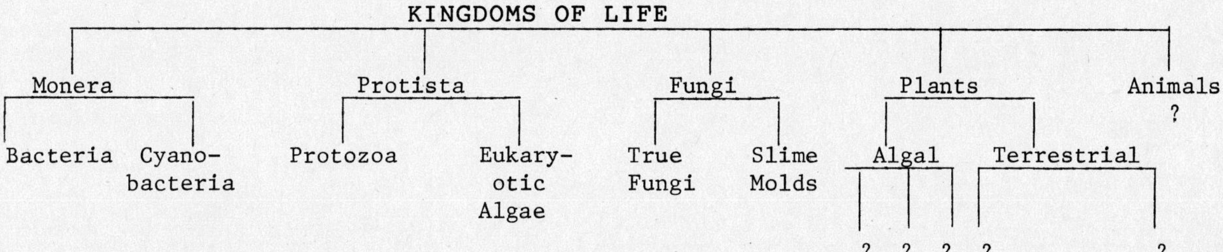

You should be able to fill in the lower level information for the three kingdoms of Monera, Protista, and Fungi. You will need to read Chapter 15 to find the information for Plants and Chapter 16 for Animals.

II. Note that the Plant kingdom is divided into <u>two</u> major divisions, not three. The first point in the Chapter Summary on page 297 gives this information.

III. Read through the new vocabulary cards. Note that there are, again, a lot of new terms. How will you organize these terms into groups? What mnemonic devices can you create? Will sentences, phrases, or words help you the most?

IV. Read the Introduction on page 286 and the Summary on pages 297, 298 and 299. Take the Post-Test on page 299.

DURING READING ACTIVITIES

I. As you read each section of the chapter, mark the paragraphs so you will be able to review important information WITHOUT REREADING the entire text. When you read the section on Multicellular Green Algae, you should note that the first half of the first paragraph gives important major characteristics and the second half of the paragraph and the two succeeding paragraphs give examples of these characteristics. If you mark this in your text and underline or highlight the characteristics in the first part of the paragraph and the last paragraph in the section, you will shorten your review time.

II. Be sure you circle new terms in the text and underline the definitions. As you complete reading a section, start rehearsing the new vocabulary. Use one of the mneumonic devices: create a sentence using the first letter of each term that is a part of a particular concept, make up a word using the first letter of each term (this is especially helpful if you don't need to remember the terms in any particular order), or devise a system of your own.

III. Read Table 15-1 on page 298 carefully. Make some vocabulary cards which have the Division on one side and the Characteristics enumerated on the reverse side.

AFTER READING ACTIVITIES

I. Check all of your textbook markings. This is a good review of the whole chapter. Remember, each time you review, you help strengthen the memory traces in your brain and the stronger the memory traces are, the more likely you are to recall the information when you need it.

II. Go back to the graphic organizer you started before you read this chapter. Fill in the lower level information.

III. Review the vocabulary. Be sure to practice this vocabulary in small groups of 507 terms at a time. Give yourself the best chance to learn by using good learning techniques.

IV. Answer the following questions.

Questions

1. Distinguish algal protists from plants.

2. List the three different types of algae and distinguish between them including their pigments.

3. Compare the life history of bryophytes and ferns and contrast them with the life history of seed plants.

4. What is the basic way in which filamentous algae reproduce?

5. The term which refers to the diploid stage in alternation of generations is ____________ the haploid ____________. If the forms are the same physically, they are said to be ____________. If they are different, ____________.

6. The major vascular tissues of plants are ____________ and __________.

7. The major source of vegetable (fruit) matter for human consumption is from the development of the ____________ into ripened fruit.

8. Distinguish between Monocots and Dicots.

Chapter 15 - Answers

1. Both organisms are capable of photosynthesizing their own food. Algae protists are unicellular eukaryotes whereas plants are multicellular.
2. Green - Chlorophyta - chlorophyll a and b - green
   Brown - Phaeophyta - fucoxanthin
   Red - Rhodophyta - phycoerythrin and phycocyanin
3. In bryophytes the gametophyte is most prominant in ferns. The sporophyte is the most dominant as it is in the other seed plants.
4. Separate algae strands lie side by side and exchange isogametes through a protoplasmic bridge called a conjugation tube. The process is call conjugation.
5. Sporophyte, gametophyte, isomorphyic, heteromorphic
6. Phloem, xylem
7. Flower
8. Monocots have a single seed leaf, flower parts usually in threes or sixes net-like leaf veins. Dicots have two seed leaves, flower parts in fours or fives and parallel leaf veins.

PLANTS	HIGHER ALGAE
CHLOROPHYTA	DESMIDS
SIPHONOUSE ALGAE	ISOMORPHIC
HETEROMORPHIC	ZOOSPORES
PHAEOPHYTA	FUCOXANTHIN
ALGIN	BLADE

SINGLE CELLED OR MULTICELLIAR ALGAE WHICH HAVE CELLULOSE CELL WALLS AND A SINGLE CHLOROPLAST PER CELL.

MULTICELLULAR AUTOTROPHS (PHOTOSYNTHESIZERS)

A SPECIAL TYPE OF GREEN ALGAE WITH A SCULPTERED CELL WALL.

A DIVISION OF MULTICELLULAR ALGAE CALLED THE GREEN ALGAE.

ALTERNATION OF GENERATIONS IN WHICH BOTH FORMS LOOK THE SAME.

A TYPE OF GREEN ALGAE WHICH HAS A FLATTENED BODY TYPE.

HAPLOID SPERES WHICH CAN MOVE BY FLAGELLA.

ALTERNATION OF GENERATIONS IN WHICH THE FORMS ARE DIFFERENT PHYSICALLY AND CHROMOSOMALLY.

THE PIGMENT WHICH PRODUCES THE BROWN COLOR IN BROWN ALGAE.

A DIVISION OF MULTICELLULAR MARINE ALGAE WHICH HAVE BROWN PIGMENTS AND ARE TERMED BROWN ALGAE.

LEAF-LIKE PORTION OF MARINE ALGAE.

A POLYSACCHARIDE FOUND ON INTERTIDAL ALGAE SPECIES.

STIPES

HOLDFAST

RHODOPHYTA

PHYCOERYTHRIN

PHYCOERYTHRIN

CORALLINE ALGAE

AGAR

CAREGEENDAN

BRYOPHYTA

RHIZOIDS

ALTERNATION OF GENERATIONS

SPOROPHYTE

ROOTLIKE STRUCTURE WHICH HOLDS ALGAE IN SUBSTRATE.

STEM-LIKE PORTION OF MARINE ALGAE.

THE REG PIGMENT IN RED ALGAE.

A DIVISION OF MULTICELLULAR MOSTLY MARINE ALGAE WHICH ARE SOMETIMES CALLED THE RED ALGAE.

RED ALGAE WITH SKELETONS OF CALCIUM CARBONATE.

THE BLUE PIGMENT IN RED ALGAE.

A FOOD ADDITIVE FOUND IN RED ALGAE.

A MEDIUM FOR GROWING BACTERIA EXTRACTED FROM RED ALGAE.

ROOTLIKE STRUCTURE OF BRYOPHYTES.

THE DIVISION OF NON-VASCULAR LAND PLANTS INCLUDING MOSSES AND LIVERWORTS.

DIPLOID SPORE FORMING PHASE IN ALTERNATION OF GENERATION.

THE CHARACTERISTIC LIFE CYCLE OF PLANTS WHICH ALTERS BETWEEN THE DIPLOID (USUALLY SPORE PRODUCING) STAGE, THE SPOROPHYTE AND THE HAPLOID STAGE, THE GAMETOPHYTE.

GAMETOPHYTE

TRACHEOPHYTA

VASCULAR TISSUE

POLYPHYLETIC

GYMNOSPERMS

CONIFERS

ANGIOSPERMS

FLOWERS

EVERGREEN

DECIDUOUS

MONOGOTYLEOONTE

ENOOSPERM

THE DIVISION OF MULTICELLULAR AUTOTROPH WHICH HAVE VASCULAR TISSUES.	HAPLOID SEED PRODUCING PHASE IN ALTERNATION OF GENERATIONS.
LITERALLY MEANS TO BE DERIVED FROM SEVERAL ORIGINS. AN EXAMPLE WOULD BE THE ANGIO-SPERMS AND GYMNOSPERMS.	CONDUCTING SYSTEMS OF PHLOEM AND XYLEM.
CONE BEARING GYMNOSPERMS.	"NAKED SEEDED" PLANTS, THE CYCADS AND CONIFERS.
SEXUAL ORGANS OF ANGIOSPERMS.	"PROTECTED SEEDED" PLANTS, THE FLOWERING PLANTS.
LOOSE LEAVES OCCASIONALLY.	ALWAYS WITH LEAVES.
FOOD STORAGE TISSUE.	SEED HAS BUT A SINGLE EMBRYONIC LEAF (COTYLEDON).

DICOTYLEDON

SEED HAS TWO EMBRYONIC LEAVES.

CHAPTER 16

ANIMAL LIFE

BEFORE READING ACTIVITIES

I. Review the information from Chapters 14 and 15. Remember that these three chapters--14, 15 and 16-- are all part of a unit which discusses the "Diversity of Life." Go over your graphic organizer and/or the chapter outlines and recite (to yourself or to a study partner) what each main heading means and what the lower level concepts are within each main heading.

II. Chapter 16 is a long and complex chapter. It is important to get an overview of what concepts are important BEFORE you begin reading it carefully.

III. Read through the outline below and answer the questions which follow.

Animal Life

I. What is an animal?
II. Their place in the environment
III. Animal relationships
    A. Acoelomates and coelomates
    B. Protostomes and deuterostomes
IV. Phylum Porifera (the sponges)
    A. Body plan of a sponge
    B. Life-style of a sponge
V. Phylum Cnidaria (coelenterates)
    A. Body plan of a cnidarian
    B. Life-style of a hydra
VI. Phylum Platyheiminthes (flatworms)
    A. Getting a head and other advances
    B. Life-style of a planarian
    C. Adaptations for a parasitic life-style--the flukes and tapeworms
VII. Phylum Nemertinea (proboscis worms)
VIII. Phylum Nematoda (roundworms)
IX. Phylum Mollusca
    A. Body plan of a clam
    B. Life-style of a clam
X. Phylum Annelida (segmented worms)
    A. Body plan of an earthworm
    B. Life-style of an earthworm
XI. Phylum Arthropoda (animals with jointed feet)
    A. The insects--secrets of success
    B. Life-style of a grasshopper
XII. Phylum Echinodermata (spiny-skinned animals of the sea)
    A. Life-style of a sea star
XIII. Phylum Chordata
    A. What is a vertebrate?
    B. Fish
    C. Amphibians
    D. Reptiles
    E. Birds
    F. Mammals
        1. Early mammals
        2. Modern Mammals

Focus on Symmetry and Body Plan
Focus on Adaptations of Squids and Octopods
Focus on Placental Mammals

1. How many phylum will you need to learn about?

2. What types of information will you be expected to remember about each phylum?

3. Where in the chapter will you get this information?

4. What do you ALREADY KNOW about each of the animals in a phylum? This is important because it will give you a background with which to ASSOCIATE NEW INFORMATION.

5. Which phylum seems to be most important based on the number of sections in the outline?

IV. Read the introduction AND the section titled, "What is an animal?" on page 302. Read the summary on pages 345 and 346. Also note Table 1601 on pages 340 and 341. Take the Post-Test on page 346.

V. Read through the new vocabulary cards. There are a lot of new terms. Start thinking now about which terms can be categorized (organized) into groups.

DURING READING ACTIVITIES

I. Since this chapter has many sections, it is particularly important that you read it carefully, section by section, mark it section by section, and then review it section by section. This means you should read from the major heading "Animal Relationships" to "Phylum Porifera" and then stop and review. Steps in your review might include:

1. Going through your textbook marking: highlighting and margin notes.

2. Finding the vocabulary cards for each term and making sure that you have circled the term in the text and underlined the definition.

3. Testing yourself on the vocabulary.

4. Making a graphic organizer to get a visual picture of how the terminology fits together. An example using the information on "acoelomates and coelomates" on pages 303 and 305 might be:

```
                                  gastrovascular-------->acoelomates
Coelom-------->germ layers-------->                        pseudocoelom
                                  tube-body------------->
                                                         coelomates
```

After you have made your own organizer, compare it with the phylogenetic tree on page 305. When you make a graphic organizer, construct it so that it makes sense to you.

5. This is also a good time to construct and begin using the mneumonic devices to help you remember vocabulary and concepts.

AFTER READING ACTIVITIES

I. Read this chapter from the beginning. Go over each section and rehearse the information.

II. Make study cards for the Phyla in Table 16-1.

A. Use 5" x 8" index cards.

B. One one side, write the name of the phylum.

C. On the other side write the important characteristics of the phylum.

D. Try to repeat the characteristics when you only look at the side with the name.

III. Review the vocabulary.

IV. Go back to the graphic organizer you began in Chapter 15 (see page of this text). Fill in the missing information.

V. Answer the following questions.

1. An animal which has a backbone is called a ___________, those without backbones are called _______________. The major ecological role that animals play is one of ____________ which means they cannot produce their own food.

2. The two basic types of animals with relation to symmetry are ____________ in which the animals are mostly attached, called ____________, and the other type in which animals have a distinctive head and ________________ and are free living are termed _______________ symmetrical.

3. Beginning with the outer most layer indicate the embryonic germ layers and indicate what they give rise to as adults.

4. Animals which have only one opening in their digestive system are said to have _______________ cavities; whereas, those with a _______________ _______________ cavity have a mouth and anus. Animals which have a fully lined body cavity are called _______________ or ____________, those like a roundworm which have an incomplete lining are called _______________.

5. List the major cells of the sponge and indicate what each one does.

6. The most unique cells of cnidarians are stinging cells which have thread-like devices called ____________ which are used to attack predators or prey. Cnidarians have two life forms the sessile ____________ and the free swimming ____________.

7. Distinguish between an intermediate and definative host.

8. Earthworms are ____________ which means its body is divided by ____________ into units which are basically all the same.

9. What structures are involved in earthworm movement.

10. Arthropods can be characterized by having ____________ appendages and a hard outer shell called an ____________. This outer shell is ____________ periodically so that the organism can change its form, the process is called metamorphosis.

11. Identify the major parts of a mollusk.

12. The __________ __________ __________ of echinoderms is made up of many __________ __________ which in conjunction with the muscular bulb ____________ are able to force open clams.

13. List the major chordate characteristics.

14. Animals who are able to regulate their temperature such as birds and mammals are called ________________. Those which cannot regulate their temperature are called ________________.

15. Mammals which lay eggs are called ____________, those which have pouches are called ______________ and those which have an ____________ egg are called ____________.

Chapter 16 - Answers

1. Vertebrate, invertebrates, consumer
2. Radial, sessile, cephalization, bilaterally
3. The outer most layer is ectoderm which gives rise to the body covering called the epidermis and the nervous system. The next layer is the mesoderm and gives rise to many systems such as the muscular, skeletal and circulatory systems. Lastly, to the inside is the endoderm which gives rise to the gastrodermis.
4. Gastrovascular, complete digestive, eucoelomates, coelomates, pseudocoelomates
5. The sponges are supported by spicules. The other cell which is important is the collar cell which helps to create currents which bring in food into the organism.
6. Nematocysts, medusa
7. An intermediate host is one which contains some immature life form of the parasite while the definative host houses the adult form.
8. Segmented, septa
9. The earthworm has a set of antagonistic muscles, circular and longitudinal, which alternately contract and relax thereby extending the body. The body is fluid filled and acts as a hydrostatic skeleton holding the body in the appropriate position. The organism is held in place by either setae or parapodia in the area which is tightly opporessed to its substrate. Alternate waves of contraction and relaxation move the animal forward.
10. Jointed, exoskeleton, molted
11. The majority parts of a mollusk are the shell, an outer hardened covering which is secreted by the mantle, a fleshy tissue which helps the animal respire and the visceral mass, the hump of viscera which is covered by the shell and the foot which is a muscular flattened structure on which the visceral mass sets.
12. Water vascular system, tube feet, ampulla
13. The major chordate characteristics are notochord, dorsal nerve cord and pharyngeal gill grooves.
14. Poikilotherms, homeotherms
15. Monotremes, marsupials, amniotic, placentals

VERTEBRATE	INVERTEBRATE
SESSILE	FREE LIVING
ASEXUAL	SEXUAL
ZYGOTE	CONSUMER
PRODUCER	PLANKTON
PARAZOA	EUMETAZOA

AN ANIMAL WITHOUT A BACKBONE.

AN ANIMAL WITH A BACKBONE.

AN ANIMAL WHICH IS ABLE TO MOVE AT WILL, NOT ATTACHED TO A SUBSTRATE.

AN ANIMAL WHICH IS ATTACHED TO THE SUBSTRATE.

ORGANISMS WHICH PRODUCE GAMETES, SPERM AND EGG IN THE REPRODUCTIVE PROCESS.

THE ABILITY TO REPRODUCE WITHOUT MATING BETWEEN SEXES.

ORGANISMS WHICH FEED ON A VARIETY OF ORGANISMS--CANNOT PRODUCE THEIR OWN FOOD.

THE UNION OF EGG AND SPERM TO PRODUCE A FERTILIZED EGG.

A READILY AVAILABLE SOURCE OF FOOD IN AQUATIC ENVIRONMENTS, SMALL MICROSCOPIC ORGANISMS.

ORGANISMS WHICH CAN PRODUCE THEIR OWN FOOD, PHOTOSYNTHESIZERS.

ALL ANIMALS BESIDES SPONGES.

THE "SIDE ANIMALS," A NAME GIVEN TO THE SPONGES.

RADIAL SYMMETRY

RADIATA

BILATERAL SYMMETRY

BILATERIA

COELOM

GERM LAYERS

ECTODERM

MESODERM

ENDODERM

GASTROVASCULAR CAVITY

DIGESTIVE CAVITY

ACOELOMATES

THE GROUP OF ANIMALS WHICH EXHIBIT RADIAL SYMMETRY.

ORGANISMS WHICH HAVE SIMILAR STRUCTURES ARRANGED AROUND THE BODY LIKE SPOKES FROM A CENTRAL AXIS, IF CUT ALONG THE AXIS AN INFINITE NUMBER OF MIRROR IMAGES COULD BE MADE.

THE GROUP OF ORGANISMS WHICH EXHIBIT BILATERAL SYMMETRY.

ORGANISMS WHICH CAN BE DIVIDED IN ONLY ONE PLANE TO PRODUCE MIRROR IMAGES.

THREE GERMANATIVE TISSUES FROM WHICH ALL TISSUES AND ORGANS ARE MADE.

A BODY CAVITY IN WHICH THE ANIMALS VISCERA LIE.

THE MIDDLE GERM LAYER WHICH GIVES RISE TO MUSCLE, SKELETAL AND CIRCULATORY SYSTEMS.

THE OUTSIDE GERM LAYER WHICH GIVES RISE TO THE BODY COVERING AND NERVOUS SYSTEM.

THE TYPE OF DIGESTIVE CAVITY WHICH HAS ONE OPENING.

THE INSIDE GERM LAYER WHICH GIVES RISE TO THE LINING OF THE DIGESTIVE SYSTEM AND OTHER ORGANS.

ANIMALS WHICH DO NOT HAVE A BODY CAVITY, LIKE JELLYFISH.

THE TYPE OF DIGESTIVE CAVITY WHICH IS COMPLETE WITH A MOUTH AND AN ANUS.

PSEUDOCOELOMATE

COELOMATE

BLASTOPORE

PROTOSTOMES

DEUTEROSTOMES

SPICULES

SPONGOCOEL

COLLAR CELLS

HERMAPHRODITIC

EPIDERMIS

GASTRODERMIS

MESOGLEA

THOSE ANIMALS WHICH HAVE A FULLY LINED BODY CAVITY, LIKE HUMANS.

ANIMALS WHICH HAVE A PARTIALLY LINED BODY CAVITY, LIKE ROUND-WORMS.

THE BRANCH OF ANIMALS IN WHICH THE BLASTOPORE FORMS THE MOUTH, LIKE MOLLUSKS, ANNELIDS AND ARTHROPODS.

THE FIRST INWARD INFLECTION OF CELLS OF AN EMBRYO TO PRO-DUCE AN OPENING.

SKELETAL ELEMENTS OF A SPONGE TO GIVE IT SUPPORT.

THE BRANCH OF ANIMALS IN WHICH THE BLASTOPORE FORMS THE ANUS, LIKE ECHINODERMS AND CHORDATES.

UNIQUE CELLS TO THE SPONGE WHICH HAVE A FLAGELLA TO HELP CREATE CURRENTS.

THE CENTRAL CAVITY OF A SPONGE THROUGH WHICH CURRENTS FLOW AND BRING FOOD INTO THE SPONGE.

THE OUTER COVERING OF AN ORGANISM, DERIVED FROM THE ECTODERM.

HAVING BOTH SEXES IN THE SAME BODY, THE ABILITY TO PRODUCE BOTH EGGS AND SPERM.

THE JELLY-LIKE SUBSTANCE WHICH IS IN BETWEEN THE EPIDERMIS AND THE GASTRODERMIS OF JELLYFISH.

THE INNER GASTRIC SURFACE OF AN ORGANISM, DERIVED FROM THE ENDODERM.

POLYP

MEDUSA

NEMATOCYST

NERVE NET

CEPHALIZATION

PROTONEPHRIDIA

FLAME CELL

AURICLES

INTERMEDIATE HOST

FINAL OR DEFINATIVE HOST

FOOT

SHELL

THE FREE LIVING OR SWIMMING PHASE OF A LIFE HISTORY OF A CNIDARIAN.

THE SESSILE OR ATTACHED PHASE OF A LIFE HISTORY OF MANY CNIDARIANS.

THE ARRANGEMENT OF NERVE CELLS IN CNIDARIANS WHICH RESEMBLES A HAIRNET.

A PART OF THE STINGING CELL OF CNIDARIANS WHICH ATTACHES TO PREDATORS OR PREY.

THE UNIT OF A FLATWORM WHICH IS RESPONSIBLE FOR COLLECTING WASTES.

THE DEVELOPMENT OF A HEAD END WITH SENSE ORGANS.

EXTENSIONS TO THE SIDE OF THE HEAD OF A FLATWORM WHICH SENSE THE ENVIRONMENT.

THE TERMINAL CELL IN A PROTO-NEPHRIDIA WHICH HAS A NUMBER OF FLAGELLA WHICH FLICKER LIKE A FLAME.

THE HOST IN THE LIFE HISTORY OF A PARASITE IN WHICH THE MATURE FORM RESIDES.

THE HOST IN THE LIFE HISTORY OF A PARASITE IN WHICH IMMATURE FORMS OF THE PARA-SITE ARE PRESENT.

THE HARD PART OF A MOLLUSK WHICH COVERS THE VISCERAL MASS.

THE BROAD MUSCULAR VENTRALLY FLESHY PART OF A MOLLUSK,

VISCERAL MASS

MANTLE

INCURRENT SIPHON

EXCURRENT SIPHON

OPEN CIRCULATORY SYSTEM

HEAMOCOEL

TROCHOPHORE LARVA

VELIGAR LARVA

SEGMENTATION

SEPTA

SETAE

PARAPODIA

THE SOFT FLESHY FOLD OF TISSUE WHICH SECRETES THE SHELL.

THE MASS OF INTERNAL ORGANS WHICH LIE ABOVE THE FOOT IN MOLLUSKS.

WATER OUTPUT VESSEL IN MOLLUSKS.

WATER INTAKE VESSEL IN MOLLUSKS.

A CAVITY THROUGH WHICH BLOOD FLOWS.

AN INCOMPLETE SET OF VESSELS THROUGH WHICH BLOOD FLOWS.

A TROCHOPHORE LARVA WHICH HAS DEVELOPED A SHELL.

THE LARVAL FORM OF MOLLUSKS AND SOME ANNELIDS.

THE TRANSVERSE PARTITIONS WHICH SEPARATE THE BODY OF AN EARTH-WORM INTO SEGMENTS.

A BODY DIVIDED INTO LIKE UNITS SUCH AS AN ANNELID.

PADDLESHAPED APPENDAGES WHICH STICK OUT FROM THE SIDE OF POLYCHAETE WORMS.

BRISTLE LIKE STRUCTURES WHICH STICK OUT OF THE SIDE OF AN ANNELID.

NEPHRIDIA

TYPHLOSOLE

PROSTOMA

CEREBRAL GANGLIA

SUBPHARYNGEAL GANGLION

DOUBLE VENTRAL NERVE CORD

EXOSKELETON

MOLTING

TRACHAE

OSTIA

CHELICERAE

PEDIPALPS

INCREASED SURFACE AREA IN THE DIGESTIVE SYSTEM OF ANNELIDS.

ORGANS TO RID BODY OF WASTES.

THE PAIRED GANGLIA WHICH SERVE AS A BRAIN FOR ANNELIDS.

THE FIRST SEGMENT OF AN ANNELID.

THE TYPE OF NERVOUS SYSTEM FOUND IN ANNELIDS AND ARTHROPODS.

THE FIRST GANGLION BELOW THE PHARYNX IN ANNELIDS.

THE PROCESS OF SHEDDING THE OUTERMOST COVERING ON ARTHROPODS.

THE HARD OUTER COVERING OF ARTHROPODS.

OPENINGS IN THE SIDES OF THE HEART OF ARTHROPODS WHICH ALLOWS BLOOD TO REENTER FROM THE HEAMOCOEL.

A SYSTEM OF BRANCHING AIR TUBES WHICH LEAD FROM THE OUTSIDE TO INDIVIDUAL CELLS IN ANTHROPODS.

THE SECOND PAIR OF APPENDAGES WHICH ARE MODIFIED TO PERFORM A VARIETY OF FUNCTIONS IN CHELICERATES.

FIRST PAIR OF APPENDAGES OF CHELICERATES WHICH HELP TO BRING FOOD INTO THE MOUTH.

BIRAMOUS APPENDAGE

ANTENNAE

MANDIBLES

UNIRAMOUS APPENDAGE

MALPIGHIAN TUBULES

METAMORPHOSIS

NYMPHAL STAGE

SPIRACLES

PENTARADIAL SYMMETRY

WATER VASCULAR SYSTEM

TUBE FEET

AMPULLA

SENSORY APPENDAGES ARISING IN THE HEAD REGION.

APPENDAGES WHICH HAVE TWO JOINTED APPENDAGES AT THEIR ENDS AS IN CRUSTACEANS.

APPENDAGES WHICH ARE JOINTED BUT HAVE A SINGLE SHAFT.

MOUTHPARTS MODIFIED FOR GRINDING.

CHANGE IN BODY FORM.

SMALL SLENDER TUBULES RESPONSIBLE FOR COLLECTING WASTES IN ARTHROPODS.

TINY OPENINGS IN THE EXOSKELETON WHICH ARE THE START OF THE TRACHAEL SYSTEM.

IMMATURE STAGE IN MANY INSECTS.

A NETWORK OF CHANNELS THROUGH WHICH WATER RUNS IN ECHINODERMS.

BODY PARTS ARRANGED ON FIVES OR MULTIPLES THEREOF SUCH AS STARFISH.

THE MUSCULAR BULB WHICH HELPS TO CONTRACT THE TUBE FEET.

THE TERMINAL ENDS OF THE WATER VASCULAR SYSTEM.

RAYS

NOTOCHORD

DORSAL TUBULAR NERVE CORD

PHARYNEGEAL GILL GROVES

VERTEBRAE

CRANIUM

PLACODERMS

OPERCULUM

RAY-FINNED FISH

LOBE-FINNED FISH

TADPOLES

AMNION

A LONG FLEXIBLE ROD WHICH SUPPORTS THE BACK OF VERTEBRATES.

THE ARMS OF THE STARFISH WHICH RADIATE OUTWARD FROM A CENTRAL DISC.

SLITS ALONG THE NECK REGION DURING THE DEVELOPMENT OF OR PERSISTANT IN SOME ADULTS (FISH) OR CHORDATES.

A SINGLE HOLLOW CORD THROUGH WHICH IMPULSES ARE CARRIED TO AND FROM THE BRAIN.

THE BRAINCASE.

CARTILAGINOUS BONY SEGMENTS OF THE VERTEBRAL COLUMN.

A COVERING OVER THE GILLS OF FISH.

A GROUP OF PRIMATIVE JAWED FISHES.

FISHES WHICH HAVE LOBES FOR FINS, GAVE RISE TO TETRAPOD LAND VERTEBRATES.

FISHES WHICH HAVE RAYS IN THEIR FINS.

REFERRING TO AN EGG TYPE IN WHICH THE EMBRYO IS BOUND BY A VARIETY OF MEMBRANES.

LARVAE FORM OF AMPHIBIANS.

POIKILOTHERMIC

FEATHERS

HOMEOTHERMIC

MAMMARY CLANDS

THERAPSIDS

MONOTREMES

MARSUPIALS

PLACENTALS

MODIFIED SCALES WHICH SERVE AS INSULATION AND FOR FLIGHT IN BIRDS.

ANIMALS WHICH CANNOT REGULATE THEIR TEMPERATURE.

ABLE TO PRODUCE MILK TO NUTURE THE YOUNG.

ANIMALS WHICH ARE ABLE TO REGULATE THEIR TEMPERATURE.

THE EGG-LAYING MAMMALS LIKE A DUCK-BILLED PLATYPUS.

A GROUP OF PRIMATIVE MAMMALS THOUGHT TO HAVE BEEN DERIVED FROM REPTILES.

THE YOUNG ARE ATTACHED TO THE MOTHER BY MEANS OF A PLACENTA.

THE POUCHED MAMMALS LIKE THE KANGAROO.

CHAPTER 17

THE PLANT BODY

BEFORE READING ACTIVITIES

I. Note that this is the first chapter in a new part of the textbook, "Plant Structure and Function." How does this section relate to the information in Part IV of the text?

II. Read through the outline on page 351. Answer the following questions.

1. How many major divisions are there in the chapter?

2. Which of the divisions discuss topics that you already know something about? You should already know something about tissues. You know what leaves, stems and roots are. Try to recall everything you know about each of these topics. You will be surprised by how much you already know.

III. Read the learning objectives which follow and go through the chapter to find the pages where you will get the information to fulfill each objective.

Key Term	Objectives	Pages
	After you have studied this chapter, you should be able to:	
	1. Distinguish between meristematic and permanent tissues in plants.	353, 354
	2. Describe and give the functions of the following types of permanent tissues: surface tissues, fundamental tissues, and vascular tissues.	355, 356
	3. Describe and give the function of the component parts of the roots and shoots of plants.	
	4. Describe the structure of a leaf and give the functions of the epidermis, mesophyll, and veins.	
	5. Summarize the role of guard cells in the water economy of the plant, and describe how they function and are controlled.	
	6. Describe the arrangement of vascular tissues in the stems of monocots and dicots.	
	7. Summarize the functions of phloem and xylem.	362, 363, 364, 365, 366, 367
	8. Describe the structure of a typical root.	
	9. Describe the typical respiratory adaptations of roots.	

Underline the term which tells you what you need to be able to do. Then write this term in the "Key Term" list. This is an important term because it lets you know what is expected of you.

IV. Read through the new vocabulary cards. Although there are many new terms, a lot of them are terms you are already familiar with, such as roots, leaves, pith, tip, node, blade, veins. Before you read the definitions on the back of the cards, try to give your own definition and then compare yours with the printed definition.

V. Read the Introduction and Summary Sections of the chapter. Take the Post-Test on page 374.

DURING READING ACTIVITIES

I. As you read, try to prepare your answers for each of the objectives at the beginning of the chapter. For the first objective you need to distinguish between two things. Distinguish between means, "to explain how these two things are different." Start a chart which shows how they are different. Your chart will need to have two columns, one for "meristematic" and one for "permanent" across the top or down the side. To determine how many columns you will need in the other direction, you must decide what characteristics you should include. You might use Table 17-1 as a guide and have three columns labelled, "structure, location, and function." Although there is a table in the text, you will learn more if you construct your own table.

II. Note that some of the objectives ask you to do more than one thing. For example, Objective 2 asks you to "describe" AND "give the function." It is important to read these objectives carefully so you will be able to do what you are asked to do.

III. Circle the new vocabulary and underline the definitions. Check to see if there are vocabulary cards for these terms. If there are not, make your own cards.

AFTER READING ACTIVITIES

I. Review what you have read. Use the chapter outline on page 351 to test yourself. Also, check to make sure that your textbook marking is accurate.

II. Go through the objectives and try to do what each asks you to do.

II. Review the following figures: 17-6, 17-12, 17-19, 17-20. Cover up the labels and try to give the proper label AND explain what the term means.

IV. Answer the following questions.

Questions

1. The two major parts of plants, usually considered to be organ systems of plants, are ___________ and ___________.

2. List the two major functions of roots.

3. Identify the four basic types of plant tissues and indicate the major function of each.

4. Animals exhibit __________ growth; whereas, plants exhibit __________ growth.

5. The body of plants is covered by __________ tissue, some of these cells are dead and filled with waxes termed __________.

6. Water and minerals flow upward in plants via __________; whereas carbohydrates move downward via __________.

7. The material in plants which becomes the heartwood is actually __________ filled with resin materials called __________.

8. The major components of phloem are __________ which lie along side __________ which have perforations at their ends called __________ __________.

9. Refer to Table 17-1. List the four kinds of Fundamental Tissue and indicate where they are located and what their function is.

10. Refer to Figure 17-2. From the outside top of a leave to the lower surface of the leaf list the contents.

11. Briefly explain how stomata control the diffusion of gases into and out of leaves.

12. Briefly describe how fluids move in plants.

13. Refer to Figure 17-16. Be able to identify the major plant tissues.

14. Refer to Figure 17-6, 17-7, 17-16 and 17-19. Compare and contrast the basic structure of minocots and dicots.

Chapter 17 - Answers

1. Roots, shoots
2. To hold plant firmly in the soil, to absorb water and minerals from the soil
3. Meristematic Tissue - Tissue which continues to divide and make young embryonic materials in plants.
   Surface Tissues - The outer layers of the plant, including epidermic and cork, which protect the body of the plant.
   Vascular Tissues - The conductive tissue which moves water upward (xylem) and carbohydrates downward (phloem).
   Fundamental Tissues - The bulk of the rest of plant tissue comprised mainly of parenchyma cells.
4. Determinate, indestermate
5. Epidermis, suberin
6. Xylem, phloem
7. Xylem, ligin
8. Companion cells, sieve tubes, sieve plates
9. Parenchyma - roots, stems, leaves - photosynthesis food stores
   Chlorenchyma - leaves, same stems - photosynthesis
   Collenchyma - stems, leaf stalks - support
   Sclerenchyma - hard parts - support
10. Epidermis - stomata-guard cells
    Mesophyll - palisade and spongy paraenchyma veins
    Epidermis - stomata-guard cells
11. Stomata are thought to open and close by two factors, CO_2 levels in the guard cells and ATP to active K+ uptake. Closure is under the control of a reverse of these conditions. Also, closing of stomata may be due to the plant hormone abscisic acid.
12. Phloem - pressure flow mechanism by osmotically facilitated pumping down an osmotic gradient.
    Xylem - capillary action including adhesion and cohesion.

ROOTS

SHOOT

LEAF

STEM

MERISTEMATIS TISSUE

DETERMINATE GROWTH

INDETERMINATE GROWTH

PRIMARY MERISTEM

SECONDARY MERISTEM

PERMANENT TISSUES

SURFACE TISSUE

EPIDERMIS

PLANT STRUCTURES WHICH INCLUDE THE STEM, LEAVES AND FLOWERS.

PLANT STRUCTURES WHICH ABSORB WATER FROM THE SOIL AND HOLD THE PLANT IN THE SOIL.

THE PLANT STRUCTURE WHICH CONDUCTS WATER UPWARD AND CARBOHYDRATES DOWNWARD.

THE PLANT STRUCTURE ASSOCIATED WITH THE SHOOT IN WHICH PHOTOSYNTHESIS TAKES PLACE.

THE TYPE OF GROWTH EXHIBITED IN ANIMALS IN WHICH ALL CELLS ARE PREDETERMINED AS TO WHAT THEY WILL BECOME.

A MAJOR PLANT TISSUE WHICH IS COMPRISED OF THE SMALL CELLS AT THE TERMINAL ENDS OF THE PLANT (BUDS AND ROOTS) WHICH DIVIDE AND DIFFERENTIATE INTO ALL OTHER PLANT CELLS.

THE INITIAL MERISTEM DERIVED FROM THE BUD OR ROOT.

THE TYPE OF GROWTH EXHIBITED IN PLANTS IN WHICH THE CELLS, EARLY IN THEIR DEVELOPMENT, ARE ALL THE SAME AND LATER DIFFERENTIATE INTO SPECIFIC FUNCTIONAL CELLS.

THOSE MATURE AND SPECIALIZED PLANT CELLS WHICH INCLUDE SURFACE, FUNDAMENTAL AND VASCULAR TISSUES.

THE MATURE BODY PARTS DERIVED FROM THE PRIMARY MERISTEM.

THE OUTER CELL LAYER OF ROOTS AND STEMS WHICH PROTECT THE UNDERLYING CELLS FROM DRYING, ABRASION AND DISEASE.

THOSE CELLS WHICH COVER THE OUTER PARTS OF THE PLANT.

SUBERIN

VASCULAR TISSUE

XYLEM

PHLOEM

VASCULAR CAMBIUM

COMPANION CELLS

SIEVE TUBES

SIEVE TUBES

FUNDAMENTAL TISSUE

PARENCHYMA

AXIL

BLADE

A MAJOR PLANT TISSUE WHICH IS RESPONSIBLE FOR CONDUCTING MATERIALS IN PLANTS.

A WAXY, GUMMY WATERPROOF MATERIAL IMBEDDED IN DEAD EPIDERMAL CELLS WHICH HELP PROTECT PLANT TISSUES.

A PLANT VASCULAR TISSUE WHICH CONDUCTS CARBOHYDRATES AND HORMONES DOWN A PLANT.

A PLANT VASCULAR TISSUE WHICH CONDUCTS WATER AND MINERALS UP A PLANT.

PHLOEM CELLS WHICH SECRETE SUBSTANCES INTO THE SIEVE TUBE ELEMENTS WHICH ARE ADJACENT TO THEM.

SPECIALIZED MERISTEMATIC CELLS WHICH DIFFERENTIATES INTO PHLOEM.

PERFERATED PARTITIONS WHICH ARE LOCATED AT THE ENDS OF ADJACENT SIEVE TUBES.

THE PHLOEM CELLS RESPONSIBLE FOR CONDUCTION.

THE PULPY CELLS WHICH MAKE UP ABOUT 85% OF PLANT CELLS.

A MAJOR TYPE OF PLANT TISSUE WHICH MAKES UP THE BULK OF PLANTS.

THE DORSALLY-FLATTENED PORTION OF THE LEAF WHICH TERMINATES IN A TIP.

THE ANGLE THE PETIOLE FORMS WITH THE STEM.

CUTICLE

VEINS

MESOPHYLL

STOMATA (A STOMA)

SPONGY PARENCHYMA

PALISADE PARENCHYMA

SCLERENCHYMA

VEINS

RADIAL MISCELLES

TRANSPIRATION

RADIAL MICELLES

TRANSLOCATION

THE CONDUCTING TISSUES IN A LEAF.

AN ACELLULAR SECRETION WHICH WATERPROOFS LEAVES.

MICROSCOPIC OPENINGS IN THE SURFACE OF LEAVES, MOSTLY LOWER SURFACE, WHICH CONTROL THE EXCHANGE OF GASES.

THE PHOTOSYNTHETIC TISSUE WITHIN THE LEAF, COMPRISED OF PALISADE AND SPONGY PARENCHYMA CELLS.

THE TOP LAYER OF MESOPHYLL, CELLS WHICH ARE ELONGATED AND LIE SIDE-BY-SIDE.

THE BOTTOM LAYER OF MESOPHYLL, CELLS OF WHICH ARE LOOSELY ARRANGED WITH NUMEROUS SPACES IN BETWEEN.

THE PATTERN IN A LEAF MADE BY THE CONDUCTING TISSUE WITHIN.

DEAD CELLS IMPREGNATED WITH LIGNIN WHICH ARE VERY STRONG, LIKE ON THE MIDRIB OF LEAF VEDUS.

EVAPORATIVE WATER LOSS.

FIBERS OF CELLULOSE IN THE CELL WALL OF GUARD CELLS WHICH ARE ARRANGED IN A WAY WHICH HELPS TO PULL THE GUARD CELLS APART.

THE PROCESS OF MOVEMENT OF PHOTOSYNTHETIC PRODUCTS WITHIN THE PLANT.

XYLEM ELEMENTS WHICH ARE SCLERENCHYMA AND SUPPORT THE STEM.

TRACHEIDS

VESSELS

LENTICELS

CORK

CORTEX

CORK CAMBIUM

PITH PARENCHYMA

SECONDARY XYLEM

HEART WOOD

SAPWOOD

GROWTH RINGS

ROOT HAIRS

DEAD XYLEM CELLS WHICH CONDUCT WATER PRIMARILY IN FERNS, FLOWERING PLANTS AND GYMNOSPERMS.

DEAD XYLEM CELLS WHICH CONDUCT WATER PRIMARILY IN FLOWERING PLANTS.

LAYERS OF DEAD EPIDERMAL CELLS IMPREGNATED WITH SUBERIN AND TANNIN.

SMALL OPENINGS IN CORK WHICH ALLOW GASES TO PENETRATE THE CORK LAYER.

PRODUCES THE CORK LAYER.

PRIMARY TISSUE BOUNDED ON THE OUTSIDE BY EPIDEMIS AND INSIDE BY VASCULAR TISSUE.

ANNUAL GROWTH LAYERS OF WOODY STEMS.

THE INNER CORE OF PLANT STEMS.

YOUNGER XYLEM JUST BENEATH THE VASCULAR CAMBRIN.

THE MOST INTERNAL SECONDARY XYLEM.

HAIRLIKE STRUCTURE EXTENDING FROM ROOTS, EXPANDING THE SURFACE AREA OF THE ROOTS.

CAUSED BY LESS COMPACT SPRONGWOOD AND FINER PORED SUMMERWOOD.

PECTIN

CASPARIAN STRIPS

ENDODERMIS

PNEUMATOPHORES

PETIOLE

FIBERS

WAX IMPREGNATED CELLS IN ROOTS WHICH PREINENT LATERAL PASSAGE OF WATER.

A GLUE-LIKE POLYSACCHARIDE WHICH AIDS ROOT HAIRS IN WATER UPTAKE.

AUXILLIARY PLANT ORGANS WHICH ACT LIKE ROOTS.

SINGLE LAYER OF CELLS SEPARATING THE CORTEX OF THE ROOT FROM THE OUTER STEELE.

XYLEM ELEMENTS WHICH ARE SCLERENCHYMA AND SUPPORT THE STEM.

A STEM-LIKE PORTION OF THE LEAF WHICH ATTACHES IT TO A LOCUS CALLED THE NODE.

CHAPTER 18

REPRODUCTION IN COMPLEX PLANTS

BEFORE READING ACTIVITIES

I. Review the information from Chapter 17 because the information about the plant body is closely related to information about reproduction in complex plants.

II. Read the outline on page 375. Do you think this chapter will be as complex as Chapter 17? Why or why not? What else do you already know about reproduction? (Remember Part III, The Continuity of Life: Cell Division and Genetics.) Look at Table 15-1 on page 298 to find out where this information on reproduction fits into the total concept of the plant kingdom.

III. Read the Learning Objectives on page 375. Underline the Key Words which tell you what you need to be able to do. Find the pages of the text which will give you the necessary information. Start a plan for fulfilling each objective. For example, with objective 1 you need to discuss the advantages and disadvantages of two types of reproduction so you should start a chart which has the following columns:

Type	Advantages	Disadvantages
1. Asexual		
	1.	
	2.	
2. Sexual		
	1.	

How will you set up a comparison chart for Objective 3?

IV. Read through the new vocabulary terms. Try to guess at the definitions and then read the definitions given to see how close you were.

V. Read the Introduction and Summary Sections of the chapter. Take the Post-Test.

DURING READING ACTIVITIES

I. As you read, prepare your answers for each of the objectives. When you finish a section, review it and then write your answer for the Learning Objective related to it. Check your textbook marking to see if you have included the necessary information.

II. Make a diagram which shows sexual reproduction in flowering plants. To do this, use line drawings, similar to those in Figure 18-11 (page 383) and label the parts. Some of the parts should include: whorls, receptacle, sepals, petal, corolla, stamen, anther, pistil, and all related terms.

III. Cover the words on Figure 18-11 and test yourself on what each part is.

AFTER READING ACTIVITIES

I. Check your textbook marking.

II. Go back to the Learning Objectives and recite your answers. Check your answers with what you wrote as you were reading the chapter.

II. Answer the following questions.

Questions

1. Most plants are able to reproduce in a variety of ways. ____________ which are small plant parts which will grow into an individual if separated from the parent is termed __________ reproduction. On the other hand, plants can also undergo __________ reproduction which requires the development and fusing of gametes.

2. List the advantages and disadvantages of sexual and asexual reproduction.

ADVANTAGES:

Asexual -

Sexual -

DISADVANTAGES:

Asexual -

Sexual -

3. Gymosperms (conifers) have separate reproductive structures which are called ______ _____ for male gametes and ________ ______ for female gametes. In angiosperms, the reproductive parts are frequently produced in the same plant part called a ____________ ____________

4. List the parts of a staminale and pistilate cone and indicate the products each produces.

Staminate

1.

2.

3.

4.

Pistilate

1.

2.

5. In flowering plants the reproductive parts are found in the flower. ____________ ____________ are produced in the ____________ which sits at the top of a filament. The ____________ contains the ____________.

6. Microspores in flowering plants produce two nuclei, one which is the __________ nucleus will develop the pollen tube, whereas the ___________ nucleus will fuse with the female ____________ cell.

7. The seed leaves are ________________ which are the embryonic values.

8. Briefly describe the six types of fruit described in the text.

9. The plant hormone ________________ cause fruits to ripen, whereas ____________ triggers growth and development.

Chapter 18 - Answers

1. Propagules, asexual, sexual
2. Asexual - no need to find a mate
   - can produce large numbers
   - requires only mitosis to produce

   Sexual - introduces and maintains new genetic information

   Asexual - does not incorporate new recombination of viable stock
   - usually cannot travel long distances in despersal

   Sexual - need to find a mate to complete process
3. Staminate cones, pistilate cones, perfect flower
4. Staminate:

   Microsporangia - pollen sacs

   Microspore mother cell - original cells to produce 3 by meiosis

   Milrospores - to produce 4 by mitosis

   Pollen grains (microgamesophyte) - male gamete

   Pistilate:
   1. Megasporangium - contains ovules
   2. Megagametophyte - female gamete
5. Pollen grams, anther, ovulary
6. Tube, generative, egg
7. Cotyleoons
8. True Fruit - fruit developed in ovulary

   Accessory Fruit - fruit developed by parts other than ovulary

   Simple Fruit - fruit developed from a single flower

   Aggregate Fruit - fruit developed from several pistals

   Multiple Fruit - fruit developed from a cluster of flowers

   Nuts - fruit with hard shell developed from ovulary
9. Ethylene, gibberelin

ASEXUAL

PROPAGULES

PLANT GRAFT

SEXUAL

OVULE

POLLEN GRAIN

SEED

STAMINATE CONE

PISTILATE CONE

STAMINATE CONE CONTAINS

PISTILATE CONE CONTAINS

GAMETE NUCLEI

SMALL PLANT PARTS WHICH, WHEN SEPARATED FROM THE PARENT PLANT GROWS INTO ANOTHER PLANT. ASEXUAL REPRODUCTION.

THE METHOD OF REPRODUCTION WHICH DOES NOT REQUIRE THE UNION OF MALE AND FEMALE GAMETES.

THE METHOD OF REPRODUCTION WHICH REQUIRES THE UNION OF MALE AND FEMALE GAMETES.

CONSISTS OF A CUTTING FROM ONE PLANT CALLED THE SCION AND ATTACHED TO ANOTHER PLANT CALLED THE ROOTSTOCK.

THE MALE STRUCTURE OF A PLANT IN WHICH THE MALE PLANT SPERM IS DEVELOPED.

THE FEMALE STRUCTURE OF A PLANT IN WHICH IS DEVELOPED THE YOUNG PLANT EMBRYO.

THE MALE REPRODUCTIVE STRUCTURE OF GYMONSPERMS (CONIFERS) WHERE POLLEN IS MADE.

THE YOUNG EMBRYO OF A PLANT PRODUCED BY THE UNION OF POLLEN AND OVULE.

MICROSPORANGIA--CONTAIN POLLEN SACS
MICROSPORE MOTHER CELLS--CONTAINED IN MICROSPORANGIUM AND PRODUCE FOUR MICROSPORE BY THE PROCESS OF MEIOSIS--IN TURN PRODUCE FOUR POLLEN GRAINS BY THE PROCESS OF MITOSIS
MICROGAMETE--THE SINGLE POLLEN GRAIN

THE FEMALE REPRODUCTIVE STRUCTURE OF GYMNOSPERMS (CONIFERS) WHERE THE SEEDS WILL DEVELOP.

TWO CELLS OF MALE WHICH ARE SPERM CELLS, ONE OF WHICH FERTILIZES THE MEGAGAMETOPHYTE.

MEGASPORANGIA--CONTAINS THE OVULE
OVULE--CONTAINS THE MEGAGAMETOPHYTE--THE "EGG" CELLS
MICROPYLE--OPENING IN OVULE FOR POLLEN GRAIN

POLLEN TUBE

PERFECT FLOWERS

FLOWER

RECEPTACLE

SEPALS

PETALS

STAMENS

PISTILS

STYLE

POLLEN MOTHER CELL

NECTAR

FLOWERING CONSTANCY

THE REPRODUCTIVE STRUCTURES OF ANGIOSPERMS WHICH CONTAIN BOTH MALE PARTS (STAMENS) AND FEMALE PARTS (PISTILS).

GROWS FROM MICROPYLE TO MEGA-GAMETOPHYTE IN ORDER FOR THE MALE NUCLEI TO GET TO THE "EGG" CELL.

EXPANDED END OF THE FLOWER STEM.

PARTS LAID DOWN IN WHORLS - CONCENTRIC CIRCLES.

MODIFIED LEAF-LIKE STRUCTURES, BRIGHTLY COLORED - ALL PETALS CALLED THE COROLLA.

MODIFIED LEAF-LIKE STRUCTURES, USUALLY GREEN.

AN ELONGATED STRUCTURE FOUND IN THE CENTER OF THE FLOWER WHICH HAS A SWOLLEN BASE CALLED THE OVULARY WHICH CONTAINS THE OVULES WHERE THE "EGG" CELLS DEVELOP.

LIE TO THE INSIDE OF THE COROLLA CONTAINS LONG SLENDER FILAMENT - ON TOP OF WHICH SETS THE
ANTHER - WHICH CONTAINS POLLEN SACS CALLED
MICROSPORANGIA - INSIDE WHICH IS THE
POLLEN - MALE GAMETE

POLLEN GRAINS DEVELOP WITHIN EACH POLLEN MOTHER CELL DIVIDED BY MITOSIS TO PRODUCE FOUR HAPLIOID (IN) CELLS TO DEVELOP MILROSPORES - WHICH GIVE RISE TO INDIVIDUAL POLLEN GRAIN - EACH HAVE TWO MITOTICALLY PRODUCED NUCLEI TUBE NUCLEUS - TO DEVELOP THE POLLEN TUBE AND GENERATIVE NUCLEUS TO FERTILIZE THE EGG.

A MORE SLENDER PORTION OF THE PISTEL WHICH TERMINATES IN A SLIGHTLY EXPANDED TOP CALLED THE STIGMA - A STICKY PORTION FOR THE RECEPTION OF POLLEN GRAINS.

THE TENDENCY OF BEES AND OTHER INSECTS TO VISIT FLOWERS OF A SINGLE SPECIES AND COLOR - IN TURN THIS ASSURES A HIGH DEGREE OF POLLINATION EFFICIENCY.

FLOWERS PRODUCE THIS SUGAR AS AN ATTRACTANT TO INSECTS.

OVULE CONTAINS:

FRUIT

COTYLEDONS

EPICOTYL

HYPOCOTYL

TRUE FRUIT

ACCESSORY FRUIT

SIMPLE FRUIT

AGGREGATE FRUIT

MULTIPLE FRUITS

DRY FRUIT

SCUTELLUM

THE RECEPTACLE WHICH HAS DEVELOPED FLESHY NUTRIENTS.

ONE DIPLOID (ZN) MEGASPORE MOTHER CELL WHICH DIVIDE BY MEIOSIS TO FORM FOUR HAPLOID MEGASPORES, 3 OF WHICH DISINTEGRATE THE REMAINING MEGASPORE DIVIDES NUCLEI MITOTICALLY TO PRODUCE A MEGAGAMETOPHYTE OR EMBRYO SAC WHICH HAS 7 CELLS, ONE OF WHICH BECOMES THE EGG.

THE PLANT EMBRYO ABOVE THE ATTACHMENT TO SEED LEAVES.

SEED LEAVES OF THE EMBRYO.

DEVELOPS FROM THE OVULARY.

THE PLANT EMBRYO BELOW THE ATTACHMENT TO SEED LEAVES.

MATURES FROM A FLOWER WITH A SINGLE PISTIL; e.g. CHERRY, DATE.

DEVELOPS FROM SOME OTHER FLOWER PART; e.g., ARTICHOKE.

A CLUSTER OF FLOWERS WHICH UNITE TO FORM A SINGLE FRUIT; e.g., PINEAPPLE.

MATURE FROM A FLOWER WITH MANY PISTILS; e.g., RASPBERRIES.

PROBABLY HOMOLOGOUS WITH THE COTYLEDON.

WALL OF OVULE DEVELOPS INTO A HARD COAT.

COLEORHIZA (RADICLE)

COLEOPTILE

ETHYLENE

ALLELOPATHY

GERMINATION

GERMINATION INHIBITOR

DORMANCY

GIBBERELIN

SHEATH PROTECTING EPICOTYL

SHEATH PROTECTING FUTURE ROOT

SECRETE TOXIC SUBSTANCE INTO THE SOIL WHICH PREVENT OTHER SEED-LINGS FROM DEVELOPING

A GASEOUS HORMONE PRODUCED BY PLANTS WHICH CAUSES RIPENING OF FRUIT TO TAKE PLACE.

THE TRIGGERING OF GERMINATION ONLY AFTER A THRESHOLD OF RAIN-FALL (USUALLY) IS REACHED

THE PROCESS OF BEGINNING THE GROWTH OF AN EMBRYO

THE PLANT HORMONE WHICH TRIGGERS GROWTH AND DEVELOPMENT WHEN ENVIRONMENTAL CONDITIONS ARE CORRECT

THE SUSPENSION OF GROWTH ACTIVITIES

CHAPTER 19

PLANT GROWTH, DEVELOPMENT, AND NUTRITION

BEFORE READING ACTIVITIES

I. Review the information from Chapters 17 and 18. Remember, Chapters 27, 18 and 19 are all one unit about plant structure and function. Be sure to review and study all of the vocabulary; many of the terms from Chapters 17 and 18 will be used in Chapter 19 (for example, meristems, phloem, xylem).

II. Read the chapter outline on page 388. Answer the following questions:

1. What are the four most important concepts you will need to understand?

2. How many plant hormones are there? Which is the most important hormone? How do you know this hormone is most important?

III. Read the Learning Objectives on page 388. Underline the Key Words which tell you what you need to be able to do. Did you notice that the Key Words used are, "describe, summarize, and outline?" Each of these words asks for the same type of information: the important points only; do not give details. Find the pages of the text which will give you the necessary information.

IV. Read through the vocabulary terms. Try to categorize the group terms into units that make sense. For example, the terms with "zone" in them should go together. All of the terms with "tropism" in them should be in the same group. Try to make a graphic organizer which will help you see how the terms with tropism are related to each other.

V. Read the introduction on page 389 and the Summary on page 407. Take the PostTest on page 408.

DURING READING ACTIVITIES

I. While you are reading, remember to highlight and annotate (make notes in) your text.

II. As you finish each section, see if you can fulfill the related objective. For example, when you finish the section on "Meristems" on pages 389-391, can you describe apical meristems and their derivatives? Write your answer down and then check it.

III. Be sure you can label the parts of the plant as depicted in Figure 192.

IV. Carefully read Table 191. You may want to make a graphic overview or some study cards to make sure you understand what each of these hormones does and how they are related. Your graphic overview might have a diagram of a flower with the appropriate hormone at each target (auxin for roots, and young stems).

V. Read Figure 1917. First try to figure out what the pictorial portion depicts then read the verbal portion. It would be a good idea to create a diagram of your own (See Learning Objective 9).

AFTER READING ACTIVITIES

I. Go back to the Learning Objectives and see if you can give the appropriate answers. If you can't, review the appropriate portion of the text. Be sure that your textbook marking is accurate and complete.

II. Review the vocabulary terms from Chapter 19. Also review the vocabulary from Chapters 17 and 18.

III. Answer the following questions.

Questions

1. Distinguish between primary meristem and secondary meristem.

2. The place on a stem to look for __________ buds is just above last year's growth which is designated by a __________ scar.

3. From the active cell dividing portion of a root or terminal bud called the __________ region, the next area is the zone of __________, an area of growth and followed by a zone of __________, an area of cell differentiation.

4. The __________ cambium is a layer of secondary meristem, which adds __________ to a plant part, gives rise to __________ __________ on the inside ending up as __________ and __________ __________ to the outside.

5. A root has a positive __________ as far as gravity is concerned and a __________ tropism in reference to light. The plant hormone called __________ acid accumulates on the lower sides of roots where it __________ growth so the root grows downward not upward.

6. Describe the flowering process as it relates to photoperiod and phytochromes.

7. Refer to Table 19-1. List the five known plant hormones and speculated hormone and indicate how they act on plants.

8. Describe the three ways in which water and minerals, called the __________ solution, can enter the root.

Chapter 19 Answers

1. Primary meristem is the area of a plant, root tip or terminal bud where the majority of mitosis takes place. Secondary meristem is the area of the stem or root where lateral branches occur.
2. Axillary, leaf
3. Meristematic elongation, maturation
4. Vascular, thickness, secondary xylem, heartwood, secondary, phloem
5. Geotropic, negative, absisic, inhibits
6. Short Day Plant - is really a long night (uninterruped light) Plant - (PR) Phytochrome Red is necessary for flowering
   Long Day Plants - really a short-night plant - (PFR) Phytochrome for Red is necessary for flowering
7. Auxin - growth
   Gibberelin - elongation of cells
   Ethylene - ripening of fruit
   Cytokinin - stimulates mitosis
   Absisic acid - dormancy, root growth
   Florigen - flowering
8. Enter by active absorption inhibition through the epidermis and mycorrhizae

ZONE OF ELONGATION

ZONE OF MATURATION

MERISTEM

TERMINAL BUD

PRIMARY (APICAL) MERISTEM

SECONDARY MERISTEM

LEAF SCAR

AUXILLARY BUD

LEAF PROMORDIA

ABSCISSION LAYER

ABSCISION

SECONDARY GROWTH

AREA OF CELL AND TISSUE DIFFERENTIATION OF THE ROOT JUST ABOVE THE ZONE OF ELONGATION, ROOT HAIRS DEVELOP IN THIS AREA.

AREA OF GROWTH OF THE ROOT, JUST ABOVE THE APICAL MERISTEM OR AREA OF CELL DIVISION.

THE ORIGINAL GROWING END OF THE SEEDLING, WHICH CONTAINS ALL OF THE DIVIDING CELLS.

THE AREA IN PLANTS OF CELL DIVISION (MITOUS) WITHOUT GROWTH.

AT NODES WHICH ARE REGULARLY SPACED ALONG THE STEM, OR IN REGULARLY ARRANGED AREAS ON THE ROOTS WHERE LATERAL ROOTS DEVELOP SOME CELL DIVISION TAKES PLACE.

THE ORIGINAL AND MOST PREVALENT AREA WHERE PLANT CELL DIVISION TAKES PLACE--THE ROOT END AND TERMINAL BUD.

THE AREA JUST ABOVE THE LEAF SCAR WHERE THE NEW SEASONS GROWTH WILL BEGIN.

AREA ON STEM WHERE LEAF BREAKS AWAY WHEN THE LEAF FALLS.

AT THE BASE OF THE LEAF STEM (PETIOLE) A TISSUE LAYER WHERE THE LEAF WILL FALL OFF.

EMBRYONIC LEAVES WITHIN THE AUXILLARY BUD WHICH ARE NOT WELL DEVELOPED. APICAL PRIMARY MERISTEM TISSUE LIES IN THE CENTER OF THE LEAF PRIMIORDIA.

THE AREA OF CELL AND TISSUE DIFFERENTIATION OR ZONE OF MATURATION, INCREASES THICKNESS.

PROCESS OF LOSING LEAVES, FLOWERS OR OTHER PARTS.

PROVASCULAR CYLINDER

GROUND TISSUE

PROTODERM

VASCULAR CAMBIUM

SECONDARY XYLEM

SECONDARY PHLOEM

CORD CAMBIUM

PRIMARY THICKENING MERISTEM

PULVINUS

TROPISMS

POSATINE TROPISM

NEGATIVE TROPISM

A TISSUE LAYER WHICH BECOMES PITH.

THE CENTRAL AREA OF A STEM WHICH GIVES RISE TO PHLOEM AND XYLEM.

A LAYER OF SECONDARY MERISTEM WHICH DEVELOPS SECONDARY BASCULAR TISSUES.

A TISSUE LAYER WHICH BECOMES THE EPIDERMOUS.

PRODUCED BY THE VASCULAR CAMBIUM GRONES OUTWARD TOWARD THE EXTERIOR.

PRODUCED BY THE VASCULAR CAMBIUM GROWS TOWARD STEM OR TRUNK, OLDER SECONDARY XYLEM BECOMES SOLID HEART WOOD.

OCCURS IN MONOCOTS BEHIND THE OPTICAL MERISTEM WHICH LAYS DOWN VASCULAR TISSUE.

A LAYER OF SECONDARY MERISTEM WHICH LAYS DOWN CORK CELLS TO FORM BARK.

PLANT GROWTH RESPONSES, FOR INSTANCE TOWARD
LIGHT - PHOTOTROPISM
WATER - HYDROTROPISM
GRAVITY - GEDIROPISM
TOUCH - THIGMOTROPISM

SPECIAL CELLS AT THE BASE OF SOME LEAVES, WHICH ARE STIMULATED BY TOUCH AND LOSE THEIR TURGOR PRESSURE CAUSING THE LEAF TO COLLAPSE.

RESPONSE AWAY FROM SOME STIMULI.

RESPONSE TO SOME STIMULUS.

AUXIN

REACTION WOOD

GIBBERELIN

ETHYLENE

CYTOKININS

ABSCISIC ACID

FLORIGEN

PHOTO PERIOD

SHORT DAY PLANTS

LONG DAY PLANTS

PHYTOCHROME

SOIL SOLUTION

WOOD WHICH HAS LOTS OF AUXINS IN RESPONSE TO STRESS, FOR INSTANCE WOUND HEALING.

A PLANT HORMONE WHICH CAUSES GROWTH TO OCCUR BY REDUCING Ph WITHIN THE CELLS.

A NATURALLY PRODUCED GAS FROM PLANTS WHICH CAUSES THE RIPENING OF FRUIT. ALSO CAN HELP COMMUNICATE BETWEEN PLANTS ABOUT INSECT DAMAGE.

CAUSES PLANTS TO ELONGATE AND GET TALLER.

HELPS IN STOMATE ACTION AS PREVIOUSLY STATED. INDUCES DORMANCY IN PLANTS AND FUNCTIONS IN PLANT GROWTH.

A SUBSTANCE WHICH CAUSES RAPID MITOSIS IN MERISTAMATIC CELLS TO OCCUR, ALSO WOUND HEALING.

A PERIOD OF LIGHT NECESSARY TO TRIGGER A SPECIFIC BIOLOGICAL FUNCTION.

A YET UNDETERMINED HORMONE THOUGHT TO BE RESPONSIBLE FOR FLOWERING.

PLANTS WHICH REQUIRE A PHOTOPERIOD ABOVE A CERTAIN CRITICAL PERIOD IN ORDER TO FLOWER - A SHORT NIGHT PLANT.

PLANTS WHICH REQUIRE A PHOTOPERIOD BELOW A CERTAIN CRITICAL PERIOD IN ORDER TO FLOWER, A LONG NIGHT PLANT.

WATER IN THE SOIL WHICH DISSOLVES MINERAL IONS.

THE PLANT PIGMENT WHICH IS STIMULATED BY LIGHT AND WITH LIGHT RESPONSIBLE FOR FLOWERING.

FIELD CAPACITY

APOPLAST

NODULES

LEGHEMOGLOBIN

NITRATE REDUCTASE

A ROUTE OF ABSORPTION THROUGH THE EPIDERMAL WALL OF THE ROOT.

THE MAXIMUM AMOUNT OF RATES A SOIL CAN HOLD BY CAPILLARY ACTION.

PLANT SUBSTANCE LIKE HEMOGLOBIN WHICH ACTIVELY TRANSPORTS OXYGEN.

SPECIAL SWELLINGS ON THE ROOTS OF LEGUMES WHICH FIX AND STORE NITROGEN IN THESE AREAS.

AN ENZYME WHICH REDUCES NITRATE TO USABLE AMMONIA.

CHAPTER 20

ANIMAL TISSUES, ORGANS, AND ORGAN SYSTEMS

BEFORE READING ACTIVITIES

I. Turn to page ix in your textbook. Find Chapter 20 in the contents list. You should note that this is the first chapter in Part VI of the text. All of the next 13 chapters discuss one major topic: Animal Structure and Function. It is important that you realize that each chapter is one part of the larger unit of ideas. The information in each of these chapters is closely related so you will need the vocabulary and concepts from the earlier chapters in this section to understand the later chapters.

II. Turn to page 411. Read the outline about "Animal Tissues, Organs, and Organ Systems."

A. What are the major ideas discussed in the chapter?
1. ______________ 3. ______________
2. ______________ 4. ______________

B. What are the types of tissues:
1. ______________ 3. ______________
2. ______________ 4. ______________

C. Use the outline to construct a graphic organizer which depicts the relationships between the ideas presented in the chapter.

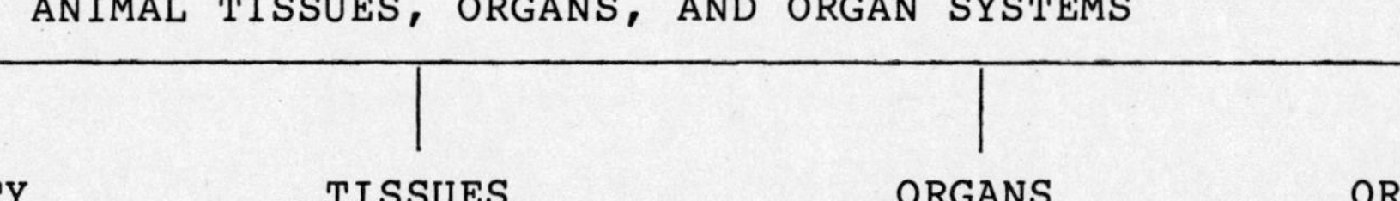

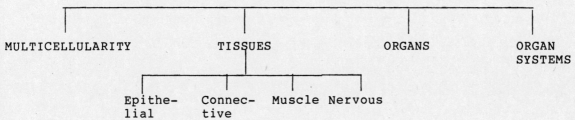

III. Read the introductory paragraph on page 412 and the summary on pages 428 and 429. Read through the vocabulary cards. Use the information from the introduction, outline, graphic organizer, summary and vocabulary cards to take the Post-Test on page 429.

IV. Vocabulary

1. Carefully read the vocabulary terms and their definitions. Sort the terms into groups that fit together.

2. You should be able to put the terms with the word "tissue" into one group without reading the text.

3. You will probably need to read the text to be able to put all of the terms dealing with epithelial tissue together: basement membrane, simple, stratified, pseudostratified, squamous, cuboidal, columnar.

4. Try to categorize (sort) as many of the terms as you can before reading then check your organization after you read.

DURING READING ACTIVITIES

I. Underline or highlight the text as you read. Mark your text carefully so you will NOT REREAD the whole chapter again. Use these steps:

1. Read a whole paragraph through completely WITHOUT MAKING A SINGLE MARK.

2. After you have read the paragraph, decide what was important.

3. Reread the paragraph and mark your answer.

4. When you have completed reading an entire section, reread your marking. Correct any errors.

II. Be sure to read the Tables in this chapter carefully:

1. Table 20-1, Epithelial Tissue
2. Table 20-2, Connective Tissue
3. Table 20-3, Muscle Tissue
4. Table 20-4, Organ Systems

III. Figure 20-11 depicts the principal organ systems of the human body. Try to picture where these systems are in your own body. You need to read this carefully and make some study cards to review these systems. Make the study cards as follows:

1. Write the system name on one side.

2. Write the parts of the system and the functions of the system on the other side.

IV. As you complete a section, fill in the major concepts on your graphic organizer.

1. In the section tittled "Tissues," stop after you finish reading about epithelial tissue.

2. You should be able to put five divisions under epithelial.

3. Read the section on connective tissue.

4. You should have seven divisions under connective.

AFTER READING ACTIVITIES

I. Review your textbook marking carefully. Reread what you marked to be sure it still makes sense. If you have omitted necessary words, be sure you correct this.

II. Check to see if you have marked all of the vocabulary terms: circled the term and underlined or highlighted the definition.

III. Test yourself on the vocabulary. Be sure that you practice the definitions of new terms in small groups of 5-7 words. Remember, you need to practice saying or writing these definitions many times to learn them.

IV. Answer the following questions.

Questions

1. Distinguish between Unicellular (=Acellular) and Multicellular organisms.

2. List the four major types of tissue and briefly state the function of each.

3. Identify the three "layer" type of epithelial tissue.

4. Identify the three shapes of epithelial tissue cells.

5. Refer to Figure 20.7 and 20.8. Be able to identify the following structures:

   MATRIX
   CHONDROCYTE
   LACUNA
   OSTEOCYTE
   HAVERSIAN CANAL
   OSTEON
   LAMELLAE

6. ________________ muscle is attached to bones and has many ________________. Cardiac is under nervous control which is ______________. ______________ muscle is found in walls of the stomach and intestines and is under ________________ control.

7. Be able to identify the major parts of the neuron.

8. Refer to Table 20-4. Identify the organ systems which assist in getting things in and out of the body ________________, ________________, ________________, ________________. Which systems integrate body functions? ________________, ________________. Which systems are involved in movement? ________________, ________________, ________________. Which system is the longest? ________________. Which system passes on genetic information? ________________.

Chapter 20 - Answers

1. Unicellular are organisms which are comprised of single cells or colonies of cells which possess organelles (cell parts) which carry on specific functions.
   Multicellular - are organisms which undergo embryology to form tissues, organs and organ systems which are involved with specific body functions.
2. Epithelial - layers of cells which cover body surfaces or lining a cavity within the body.
   Connective - layers or structures which bind together, support and protect other tissues.
   Muscle - layers or structures specialized for contraction.
   Nervous - tissues which transmit nerve impulses.
3. Simple - one cell thick.
   Stratified - two or more cells thick in layers or stacks.
   Pseudostratified - appear to be stacked but not.
4. Squadmous- flat.
   Cuboidal - cube.
   Columnar - elongated.
5. Refer to Figure 20.7 and 20.8.
6. Skeletal, striations, involuntary, smooth, involuntary
7. Dendrite, cell body, axon, glial cells
8. Digestive, circulatory, respiratory, urinary, nervous, endocrine, muscular, skeletal, nervous, integument, reproductive

MULTICELLULAR

TISSUE

EPITHELIAL

BASEMENT MEMBRANE

SIMPLE EPITHELIUM

STRATIFIED EPITHELIUM

PSEUDOSTRATIFIED EPITHELIUM

SQUAMOUS EPITHELIUM

CUBOIDAL EPITHELIUM

COLUMNAR EPITHELIUM

MICROVILLI

GLAND

A SPECIAL GROUP OF LIKE CELLS WHICH ARE ASSOCIATED INTO A LAYER (USUALLY) TO DO A SPECIFIC FUNCTION; e.g., SKIN-EPIDERMIS FOR PROTECTION.

ORGANISMS WHICH UNDERGO DEVELOPMENT TO PRODUCE A MANY CELLED BODY, USUALLY ORGANIZED INTO TISSUES, ORGANS AND ORGAN SYSTEMS.

NONLIVING MATERIAL WHICH IS SECRETED BY THE EPITHELIAL TISSUE WHICH ATTACHES IT TO THE UNDERLYING TISSUE.

CELLS WHICH LIE CLOSE TO ONE ANOTHER IN SHEETS TO COVER THE BODY OR LINING OF CAVITIES IN THE BODY.

TWO OR MORE CELL LAYERS.

ONE CELL LAYER TISSUE.

FLATTENED CELLS.

APPEARING TO BE LAYERED BUT NOT.

ELONGATED CELLS.

CUBE-SHAPED CELLS.

A GROUP OF EPITHELIAL CELLS SPECIALIZED TO SECRETE A PRODUCT; e.g., MILK, SALIVA.

TINY CELL MEMBRANE PROJECTIONS WHICH INCREASE CELL SURFACE AREA.

GOBLET CELLS

CONNECTIVE

STROMA

INTRACELLULAR SUBSTANCE

MATRIX

FIBROBLASTS

COLLAGEN FIBER

LOOSE CONNECTIVE TISSUE

DENSE CONNECTIVE TISSUE

ELASTIC CONNECTIVE TISSUE

RETICULAR CONNECTIVE TISSUE

ADIPOSE TISSUE

TISSUE TYPE IN WHICH THE CELLS ARE ORGANIZED IN WAYS WHICH BIND TOGETHER, SUPPORT AND PROTECT OTHER TISSUES.

CELLS IN THE FORM OF A GOBLET WHICH SECRETE MUCUS.

THREADLIKE STUSTANCES OR FIBERS BETWEEN THE CELLS IN CONNECTIVE TISSUE.

A SUPPORTING FRAMEWORK OF CONNECTIVE TISSUE SURROUNDING AN ORGAN.

CONNECTIVE TISSUE CELLS WHICH SECRETE THE INTRACELLULAR FIBERS.

A THIN GEL POLYSACCHARIDE IN WHICH THE INTRACELLULAR FIBERS ARE FOUND.

THIN SHEETS OF TISSUE SURROUNDING NERVE CELLS, BLOOD VESSELS AND MUSCLES, ALSO ADIPOSE TISSUE.

PROTEINS WHICH FORM PLASTIC-LIKE FIBERS WHICH CAN STRETCH AND RETURN TO THEIR ORIGINAL SHAPE.

PARALLEL ELASTIC FIBERS SUCH AS LIGAMENTS, WALLS OF ARTERIES, ETC.

STRONG, PRIMARILY COLLAGEN FIBERS IN BUNDLES, RESISTANT TO STRESS; e.g., TENDONS.

FAT CELLS WHICH STORE AND RELEASE FAT, CUSHIONS INTERNAL ORGANS.

INTERLACING FRAMEWORK OF FIBERS TO SUPPORT ORGANS SUCH AS THE LIVER, SPLEEN AND LYMPH NODES.

CARTILAGE

CHONDROCYTES

BONE

OSTEOCYTES

CANALICULI

LAMELLAE

OSTEONS

HAVERSIAN CANALS

OSTEOCLASTS

BLOOD

LYMPH

MUSCLE TISSUE

CARTILAGE CELLS WHICH SECRETE RUBBERY MATRIX.

SKELETON OF VERTEBRATE IN EMBRYONIC STAGES AND ADULT STAGES OF SHARKS AND RAYS.

BONE CELLS WHICH SECRETE HARD MATRIX.

PRINCIPLE SKELETAL STRUCTURE.

CONCENTRIC LAYERS OF CAPILLARIES.

TINY CANALS THROUGH WHICH OSTEOCYTES ARE CONTAINED.

CENTRAL CANALS IN THE OSTEONS THROUGH WHICH CAPILLARIES AND NERVES RUN.

SPINDLE SHAPED UNITS IN BONE.

CIRCULATING TISSUE INVOLVED WITH GAS EXCHANGE.

CELLS IN BONE CAPABLE OF DISSOLVING BONY TISSUE.

PROVIDES THE BASIS OF MOVEMENT IN ANIMALS BY BEING ABLE TO CONTRACT.

COLORLESS FLUID INVOLVED WITH IMMUNE SYSTEM.

SKELETAL MUSCLE

STRIATED MUSCLE

CARDIAC MUSCLE

INTERCALATED DISKS

SMOOTH MUSCLE

NERVOUS TISSUE

NEURON

GLIAL CELL

DENDRITE

AXON

ORGAN

ORGAN SYSTEM

MUSCLE WHICH HAS LIGHT AND DARK AREAS DUE TO THE INTERNAL STRUCTURES.

VOLUNTARY MUSCLES ATTACHED TO THE BONE.

MARK THE JUNCTION BETWEEN MUSCLE FIBERS.

HEART MUSCLE UNDER INVOLUNTARY CONTROL.

SPECIALIZED CELLS TO RECEIVE AND TRANSMIT INFORMATION IN TURN CONTROL AND INTEGRATE BODILY FUNCTIONS.

INVOLUNTARY MUSCLES WHICH LACK STRIVATIONS FOUND IN THE WALLS OF MANY ORGANS.

A SUPPORTING CELL SURROUNDING THE NEURONS.

THE FUNCTIONAL CELL OF NERVOUS TISSUE.

NERVE CELL PART WHICH CONDUCTS IMPULSES AWAY FROM THE CELL TO ANOTHER CELL OR BODY PART.

NERVE CELL PART WHICH RECEIVES IMPULSES.

A COMBINATION OF VARIOUS TISSUES AND ORGANS WHICH DO A SPECIFIC BIOLOGICAL FUNCTION.

A GROUP OF TWO OR MORE KINDS OF TISSUES ORGANIZED TO DO A SPECIFIC BIOLOGICAL FUNCTION.

ORGANISM

INTEGUMENT

SKELETAL

MUSCULAR

DIGESTIVE

CIRCULATORY

RESPIRATORY

URINARY

NERVOUS

ENDOCRINE

REPRODUCTIVE

ACTS AS A BARRIER TO THE ENVIRONMENT, HELPS TO CONTROL TEMPERATURE AND PROTECTS THE BODY.

A MULTICELLULAR FORM COMPRISED OR ORGAN SYSTEMS.

MOVES BODY PARTS AND INTERNAL MATERIALS

SUPPORTS THE BODY, PROVIDES FOR MUSCLE ATTACHMENT FOR MOVEMENT.

TRANSPORT GASES AND FOOD TO SITES OF CELLULAR RESPIRATION.

PROVIDES FOR THE BREAKDOWN AND ABSORPTION OF ENERGY SOURCES.

RIDS BODY OF METABOLIC WASTES AND BALANCES IONS AND WATER.

EXCHANGE OF GASES BETWEEN ATMOSPHERE AND CIRCULATING.

REGULATES BODY CHEMICALS WHICH ACT ON VARIOUS BODY FUNCTIONS.

RECEIVES, CONDUCTS AND CONDUCTS IMPULSES THEREBY INTEGRATING THE TOTAL BODY.

PRODUCES GAMETES FOR SEXUAL REPRODUCTION.

CHAPTER 21

SKIN, MUSCLE, AND BONE

BEFORE READING ACTIVITIES

I. Before you read Chapter 21, review Chapter 20. Use your graphic organizer to review the major concepts.

1. What are the advantages of multicellularity?

2. Describe each type of tissue. How many types of epithelial tissue? What are they? How may types of connective tissue? Muscle Tissue?

3. What are the organs?

4. What are the organ systems? Where is each one? What does each one do?

II. Read the outline below and answer the questions which follow it.

Skin, Muscle, and Bone
- I. The vertebrate skin
  - A. The epidermis
  - B. The dermis
- II. Skeletons
  - A. Hydrostatic skeletons
  - B. External skeletons
  - C. Internal skeletons
- III. Muscle
  - A. Muscle structure
  - B. The biochemistry of muscle action
  - C. Variations in muscle response

Focus on molting
Focus on building bone
Focus on flight muscles of insects

1. What are the three most important sections of this chapter?

   1. ____________________ 3. ____________________
   2. ____________________

2. What are the skin types? How do you think they differ?

3. What are the types of skeletons? How do you think these are different? what type of animal do you expect to have each type of skeleton?

4. What are the important concepts you need to learn about muscles? What do you already know about biochemistry? (Think about what you learned in part I of the text, The Organization of life.)

III. Use the vocabulary sheet to become familiar with the new terminology used in this chapter.

1. Separate the sheet into individual cards.

2. Read the word on the fron and GUESS what the definition may be. check to see how close your guess is to the actual definition.

3. Sort the terms into groups which seem to fit into the major sections of the chapter.

4. Read the introduction, summary, and vocabulary terms. Take the Post-Test on pages 450 and 451.

DURING READING ACTIVITIES

I. Underline or highlight the text as you read. Remember to read the whole paragraph before you make any marks. After you highlight or underline, you should summarize the meaning in the margin. The example below is from page 433.

The Biochemistry of Muscle Action

The interaction of the actin and myosin filaments produces the contraction of muscle. Each myosin filament consists of about 200 molecules of the protein myosin in a parallel arrangement. A rounded head extends from each rod-shaped myosin molecule. The head of the myosin molecule bears a binding site that is complementary to binding sites on the actin filament. Each actin filament contains 300 to 400 rounded actin molecules arranged in two chains.

A unit of actin and myosin filaments makes up a sarcomere. (Fig. 21-13). There are many sarcomeres in a skeletal muscle cell, united at their ends by a comples interweaving of filaments called the Z line. Thus, the Z line indicates the boundary of each sarcomere. Each sarcomere is capable of independent contraction. When many sarcomeres contract together, they produce the contraction of the muscle.

1. Note that only key words are used to make the most important points stand out.

2. Most of the first paragraph explains how actin and myosin filaments interact. This is a detail that is not as important as the major idea that it is the combination of two types of filaments, actin and myosin, that work together to cause a muscle to contract.

3. The second paragraph adds to the information in the first paragraph by telling how the filaments are organized so they can work together to cause a muscle contraction.

4. If you sum up this information in the margin, you will NOT need to REREAD these paragraphs. Be sure that you explain the the important concepts in your won words.

II. Be sure that you read the figures and think about how they explain the verbal portion of the text.

1. Use Figure 21-13 to review the information you learned in the two paragraphs above. Try to explain the diagrams in your own words.

2. Other figures you should be sure to read carefully include 21-1, 21-14, and 21-16.

III. Use your vocabulary cards as you read.

1. Check to see if you have vocabulary cards for each new term you find.

2. Circle the term in the textbook and find the vocabulary card that defines it.

3. Try to define the term BEFORE you read the definition.

4. If you find terms which are new to you and not on the vocabulary cards, use the blank cards to add these terms to your set.

AFTER READING ACTIVITIES

I. Review your textbook marking. Be sure that you understand what you marked and why it is important. If you omitted somthing important, add it now.

II. Review the vocabulary. Study the terms in groups which go together. You should group the terms about skin together, the terms about skeletons together, and the terms about muscles together.

III. Answer the question which follow.

1. The largest single organ in the human body is the ____________. This outer unit has a variety of inclusions which do specific functions. For instance ____________ glands help the body help the body cool down. Some modified structures for special acitvity are ____________ for birds, ____________ for mammals and ____________ for reptiles.

2. Distinguish between the various layers of skin.

3. Briefly define and distinguish between the three major support systems in animals and give an example of each.

4. Briefly explain how an earthworm moves.

5. The problem with an exoskeleton is that the rigid structure prevents _____________; therefore, the organism must ___________ in order to grow.

6. The molting process is initiated by ___________ which in turn does two things. It initiates the development of a new ___________ and produces an enzyme which breaks down the old ___________.

7. What are the advantages of an ENDOSKELETON?

8. Distinguish between the AXION and APPENDICULAR skeletons.

9. A __________ __________ is one which orginates as ___________ and ossifies into _____________. A flat bone, or__________ __________ is one which starts as splinter like parts of bones called __________ then lays down more bone to the outside.

10. Muscle cells sometimes called ___________ consist of smaller thread like units called ___________ which in turn are made up of thick filaments _____________ and their filaments ____________. These two types of filaments have structures which bind the two together termed ___________ ___________.

11. Briefly describe how muscles contract.

Chapter 21 - Answers

1. Skin, sweat, feathers, hair, scales
2. Epidermis - the outermost layer - comprised of two sublayers. Stratum Basale which create new cells and Stratum Corneum which become hardened with keratin and slough off the body.
   Dermis - the innermost layer made of collagen fibers
3. Hydrostatic Skeleton - a support system created by a fluid filled cavity which has a different pressure within the system than ouside the system such as found in roundworms and earthworms.
   Exoskeleton - a support system and site for muscle attachment which is on the outside of the body such as the arthropods and mollusks.
   Endoskeleton - a support system and site for muscle attachment which is on the inside of the body such as echinoderms and chordates.
4. Earthworms have an hydrostatic skeleton in which each of the segments, partitioned by septa, acts as an independent unit. Their fluid filled segments and the alternating contraction of circular and longitudinal muscles in a "wave-like" fashion are the primary features of earthworms movement. Additionally, the thickening of the worm in the burrow allows it to anchor to the burrow as the segments ahead move forward and attach themselves in a similar fashion. The process is repeated time and again, speed of reaction is due to giant nerve axons.
5. Growth, molt
6. Ecdysone, cuticle, cuticle
7. An internal skeleton provides protection, support and attachment of muscles for movement.
8. Axial - the parts of the skeletal axis.
   Appendicular - the appendages and their attachments.
9. Cartilage bone, cartilage, bone, membrane bone, spicules
10. Fibers, myofibrils, myosin, actin, cross bridges
11. After nerve cells stimulate a muscle to contract the muscle cell membrane undergoes depolarization causing the release of calcium ions which in turn causes the actin and myosin fibers to slide on each other. ATP is required for this contraction to occur, the source of the energy-rich phosphate is from creatine phosphate.

SKIN

SEBUM

ACNE

EPIDERMIS

KERATIN

DERMIS

EXOSKELETON

HYDROSTATIC SKELETON

ANTAGONISTIC MUSCLES

SEPTA

EXTERNAL SKELETON

CUTICLE

A WAX SUBSTANCE WHICH IS SECRETED BY OIL GLANDS, COVER FACE AND SCALP.

THE OUTER SURFACE OF ANIMALS WHICH IS FREQUENTLY MADE UP OF SENSORY RECEPTORS, GLANDS, AND MODIFIED STRUCTURES SUCH AS HAIR AND SCALES.

THE OUTER LAYER OF SKIN WHICH IS FURTHER DIVIDED INTO
STRATUM BASALE - CLOSEST TO THE BODY, UNDERGO DIVISION
STRATUM CORNEUM - OUTERMOST LAYER, BECOME HARDENED.

INFLAMMATION OF THE FACE CAUSED BY EXCESSIVE SEBUM IN CONJUNCTION WITH THE ELEVATION OF SEX HORMONES.

THE LAYER TO THE INSIDE OF EPIDERMIS, CONSISTS OF COLLAGEN FIBERS.

A PROTEIN IN THE EPIDERMIS WHICH WATERPROOFS AND STRENGTHENS OUT OUTER CELLS OF THE SKIN.

A FLUID FILLED CAVITY WHICH ACTS AS A STRUCTURAL SUPPORT SYSTEM IN SOFT BODIED ANIMALS LIKE EARTHWORMS.

DEAD TISSUE WHICH CREATES A SHELL OR HARDENED OUTER COVERING SUCH AS ON A GRASS-HOPPER.

INTERNAL PARTITIONS IN AN EARTH-WORM WHICH DIVIDE IT INTO MANY SIMILAR COMPARTMENTS WHICH ARE ALL FLUID FILLED.

PAIRS OF MUSCLES WHICH OPERATE IN OPPOSITION TO ONE ANOTHER.

THE NONLIVING LAYER OF ARTHROPOD EXOSKELETON TO THE OUTSIDE OF THE EPIDERMIS.

SUPPORT AND MUSCLE ATTACHMENT IS ON THE OUTSIDE OF THE ORGANISM e.g., MOLLUSKS AND ARTHROPODS.

CHITIN

EPICUTICLE

INSTARS

ECOYSONE

ECDYSIS

ENDOSKELETON

AXIAL SKELETON

APPENDICULAR SKELETON

ARTICULAR CARTILAGES

ARTHRITIS

JOINT CAPSULES

SYNOVIAL FLUID

THE OUTER MOST PART OF THE EXOSKELETON WHICH IS A PROTECTIVE WAXY SUBSTANCE.

A LONG CHAIN SUGAR SIMILAR TO CELLULOSE WHICH IS THE BASIC SUBSTANCE OF ARTHROPOD EXOSKELETONS.

A HORMONE WHICH INITIATES THE MOLTING PROCESS.

DEVELOPMENTAL STAGES BETWEEN MOLTS OF AN INSECT.

LIVING TISSUE WHICH IS INTERNAL; e.g., ECHINODERMS AND CHORDATES.

THE PROCESS OF SHEDDING THE OUTER LAYER OF THE EXOSKELETON TO ALLOW GROWTH TO OCCUR.

THE SKELETON WHICH INCLUDE THE APPENDAGES AND THEIR ATTACHMENTS, THE SHOULDER AND PELVIC GIRDLE.

ENDOSKELETON NEAR THE AXIS OF THE ORGANISM; e.g., THE SKULL, SPINE, RIBS AND STERNUM.

DISEASE OF THE BONES WHERE CALCIUM DEPOSITS FORM.

CONNECTIVE TISSUE AT THE END OF BONES WHICH ACT AS BEARINGS BETWEEN THE BONES.

LUBRICANT IN THE JOINT CAPSULES.

FLUID FILLED JOINTS AT THE ENDS OF BONES.

CARTILAGE BONE

COMPACT BONE

CANCELLOUS BONE

MARROW CAVITY

PERIOSTEUM

MEMBRANE BONE

SPICULES

FIBERS

MYROFIBRILS

MYOFILAMENTS

MYOSIN

ACTIN

OUTER COVERING OF VERY DENSE BONE MATERIAL.

THE NAME GIVEN TO SKELETAL ELEMENTS WHICH ORIGINATED AS CARTILAGE AND OSIFIED INTO BONE.

INTERIOR CAVITY IN BONE.

LIES BENEATH THE COMPACT BONE, SPONGY IN NATURE.

A BONE PRODUCED FROM EMBRYONIC MEMBRANE OF CONNECTIVE TISSUE; e.g., FRONTAL BONE ON SKULL.

A THIN MEMBRANE WHICH COVERS AND LAYS DOWN LAYERS OF BONE.

ELONGATED MUSCLE CELLS.

THE FORMATION OF SPLINTER-LIKE STRUCTURES WHICH ARE THE BEGINNING OF MEMBRANE BONE FORMATION.

EVEN SMALLER COMPONENTS OF THE MUSCLE STRUCTURE WHICH ARE:

SMALLER UNITS IN MUSCLE FIBERS, THREAD LIKE, WHICH RUNS THE LENGTH OF THE CELL.

THIN PROTEIN

THICK PROTEIN

SARCOMERE	Z-LINE
ACETHYLCHOLINE	MYONEURAL CLEFF
SARCOLEMMA	DEPOLARIZATION
T-SYSTEM	SARCOPLASMIC RETICULUM
CROSS BRIDGES	FLIGHT MUSCLES
PILUS	GLYCOGEN

THE BOUNDARY OF SARCOMERES.

A UNIT OF ACTIN AND A UNIT OF MYOSIN.

THE SPACE BETWEEN THE NERVE ENDING AND THE MUSCLE WHERE ACETYLOCHLINE WILL INITIATE THE CONTRACTION OF THE TISSUE.

A CHEMICAL SIGNAL WHICH CAUSES CONTRACTION OF MUSCLE.

THE CHANGE OF ELECTRICAL CHARGE AT THE SURFACE OF THE MUSCLE CELL.

THE CELL MEMBRANE COVERING THE MUSCLE CELL.

SPECIALIZED TYPE OF ENDOPLASMIC RETICULUM, RELEASES CALCIUM, IMPORTANT IN DEPLORIZATION.

THE INWARD EXTENSION OF THE CELL MEMBRANE OF A MUSCLE CELL.

STRIATED MUSCLE IN INSECTS.

ENDS OF MYOSIN MOLECULES WHICH BRIDGE THE GAP BETWEEN THE THICK AND THIN FILAMENTS.

A LONG CHAIN POLYSACCHARIDE FROM MANY GLUCOSE MOLECULES, IMPORTANT IN MUSCLE ACTION.

INSULATING MATERIAL FOUND COVERING INSECT BODIES.

CREATINE PHOSPHATE

SIMPLE TWITCH

TETANUS

SLOW TWITCH

FAST TWITCH

A SINGLE RAPID SKELETAL MUSCLE CONTRACTION DUE TO A SINGLE BRIEF STIMULUS.

THE ORGANIC MOLECULE WHICH PROVIDE ATP THE SOURCE OF HIGH-ENERGY PHOSPHATE.

SLOW MUSCLE RESPONSE.

A SERIES OF SKELETAL MUSCLE CONTRACTIONS, WHICH BECOMES A SMOOTH SUSTAINED CONTRACTION, DUE TO A SERIES OF SEPARATE STIMULI PRODUCED IN RAPID SUCCESSION.

FAST MUSCLE RESPONSE.

CHAPTER 22

RESPONSIVENESS: NEURAL CONTROL

BEFORE READING ACTIVITIES

I. Before you begin to read Chapter 22, review the highlights of Chapter 21.

1. Explain the difference between the layers of the vertebrate skin.

2. What are the three types of skeletons and how do they differ from each other?

3. How is muscle structured?

4. What type of chemistry causes muscle action?

II. Read through the outline below and answer the questions which follow it.

RESPONSIVENESS: NEURAL CONTROL
- I. Cells of the nervous system
  - A. Gilial cells
  - B. Neurons
- II. Information flow through the nervous system
  - A. Neural circuits
  - B. Reflex action
- III. Transmission of impulses
  - A. Transmission along a neuron
  - B. Substances that affect excitability
  - C. Transmission between neurons
  - D. Neurotransmitters
  - E. Direction and speed of conduction
- IV. Neural integration
- V. Organization of neural circuits

  Focus on regeneration of an injured neuron

1. What do you already know or think about your own nervous system?

2. How do you think the information in this chapter might help you, personally?

3. What are the cells of the nervous system? Turn back to Chapter 20 and review the information about the cells of the nervous system. What are the other types of tissue in the human body?

4. How does information flow through the nervous system?

5. What concepts will you need to understand to be able to explain the transmission of impulses?

III. Read the vocabulary cards. Group them into units that make sense. Some things to consider when you are grouping vocabulary include:

1. Looking at the words that have some common word parts and trying to decide how they may go together. Some examples of terms that can be in one group include:

1. Words with glia, such as glial cells, neuroglia.
2. Words with polarization, such as depolarization, repolarization.

2. Being aware of words which you already know a meaning for, such as "resting potential", "threshold", "all-or-none." Remember that these terms have specialized meanings in this context.

IV. Read the introduction on page 453, the summary on pages 468 and 469. Take the Post-Test on pages 469 and 470.

DURING READING ACTIVITIES

I. Remember to read the chapter from one major section to the next, marking each section paragraph by paragraph, and stopping at the end of each major section to review what you have read.

II. Use the vocabulary cards as you read. Check to see if the groups you put together before you started to read make sense. If not, regroup so they do make sense.

III. Be sure that you underline or highlight as you read. After you have done this, summarize the information in your own words in the margin.

1. The example below is from page 467 of the textbook. Check your own marking to see if it is like this.

2. It is not necessary to mark whole sentences. Try to mark only the important words in each sentence. Omit marking minor details.

3. In this paragraph there are several steps given to explain how "convergence" works. It is important that you are able to repeat these steps without having to read them. Be sure to list them in the margin.

Organization of Neural Circuits

As illustrated by the reflex pathway discussed earlier in this chapter, neurons are organized into specific pathways, or circuits. Within a neural circuit many presynaptic neurons may converge upon a single postsynapitc neuron. In convergence, the postsynaptic neuron is controlled by signals from two or more presynaptic neurons. Figure 22 - 12. An association neuron in the spinal cord, for instance, may receive converging information from sensory neurons entering the cord, from neurons originating at other levels of the spinal cord, and even from neurons bringing information from the brain. Information from all of these converging neurons must be integrated before an action potential is generated in the association neuron and an appropriate motor neuron stimulated.

CONVERGENCE

1. Neurons circuiting
2. Circuits carry information
3. Circuits converge converging information integrated
4. Info integrated action potential

IV. Read all of the figures in the chapter very carefully.

1. Note that a figure is often divided into more than one part. Be sure that you understand what each part represents and how each part differs.

2. Practice reading Figure 22-1 until you can label all of the parts of a multipolar neuron without looking at the drawing. To help you in this practice, turn to page 470 and use the drawing in the Post-Test.

3. Be sure that you read Table 22-1 carefully. It summarizes the information about neurotransmitters.

AFTER READING ACTIVITIES

I. Review your textbook marking. Remember, one reason for marking your textbook carefully is to cut down the amount of reading you must do when you review for an examination.

II. Review the vocabulary.

1. Go through all the terms you circled during reading and match them with the vocabulary cards. Repeat the definitions to yourself and sort the cards into those you know and those you need to learn.

2. Study the vocabulary in groups of 5-7 terms at a time. Practice each group several times each day.

3. Review words from chapters 20 and 21 each day to make sure that you don't forget these terms.

III. Answer the questions which follow.

Questions

1. Distinquish between clial cells and neurons.

2. Name the three parts of a neuron.

3. The structure of the end of an axon is termed a ______________ _______, the chemical substance which bridges the ____________ _________ is called ______________ and in order to be able to turn off a chemical impulse another chemical called ____________ is produced.

4. Neurons are covered by two sheaths, the neurilemma is on the outside, the ___________ ___________ which serves as an __________ to cover the __________.

5. A __________ is a collection of AXONS wrapped in connective tissue; a __________ is a collection of cell bodies.

6. A reflex action is a fixed pattern reaction to a specific response, indicate, in order, what is involved in such a response.

7. When the inner surface of the cell membrane is negatively charged with the neuron is said to be in a __________ __________, when the polarity reverses for a short period of time and a stimulus reaches an appropriate ____________ an __________ __________ is reached. When a WAVE OF ____________ takes place an electric current travels down a neuron.

8. A group of associated chemicals called ____________ are responsible for changing electrical impulses into chemical impulses to keep the impulse moving to the target site. The most common of these is acetylcholine another one which is known to affect moods is _______.

Chapter 22 - Answers

1. Glial Cells are in far greater numbers than are neurons, about 10:1. neurons transmit impulses along specific pathways, Glial cells have an illdefined role frequently surround the neuron.
2. The cell body, the axon and the dendrite.
3. Synaptic knob, synaptic cleft, acetylcholine, cholinesterase
4. Myelin sheath, insulation, axon
5. Nerve ganglion
6. The reflex action or arc involves receiving a stimulus from a receptor which moves the impulse toward the spinal cord by means of a sensory neuron which transmits the impulse to the association neuron which transmits the impulse to the motor neuron which ends in conducting a message for a group of muscles to respond accordingly. All of this takes place without the impulse going to the brain.
7. Resting potential, threshold, action potential, deplorization
8. Neurotransmitters, norepinephrine

STIMULI

GLIAL CELL

NEUROGLIA

SCHWANN CELLS

NEURON

CELL BODY

DENDRITE

AXON

SYNAPTIC KNOBS

COLLATERALS

CELLULAR SHEATH

MYELIN SHEATH

A TYPE OF CELL FOUND IN THE NERVOUS SYSTEM, THE EXACT FUNCTION IS NOT KNOWN, THOUGHT TO PLAY A ROLE INSULATING NEURONS.

ANY CHANGE IN THE BODY OR OUTSIDE THE BODY THAT THE ORGANISM CAN DETECT.

A SPECIALIZED KIND OF SUPPORT AND INSULATING CELL SURROUNDING SOME NEURONS.

A COLLECTION OF GLIAL CELLS.

LARGEST PORTION OF THE NEURON, CONTAINS THE NUCLEUS.

THE FUNCTIONAL UNIT OF THE NERVOUS SYSTEM, RECEIVES AND TRANSMITS NERVE IMPULSES.

A LONG, SINGLE FIBER WHICH CONDUCTS NEURAL MESSAGES TO ANOTHER NEURON, MUSCLE OR GLAND.

A SHORT, HIGHLY BRANCHED FIBER WHICH RECEIVES NEURAL IMPULSES AND SEND THEM TO THE CELL BODY.

BRANCHES OFF THE AXON.

TINY EXPANDED ENDS OF THE AXON WHICH RELEASE CHEMICALS THAT ARE IMPORTANT IN THE TRANSMISSION OF IMPULSES FROM ONE NEURON TO THE NEXT.

THE INNER SHEATH COVERING AXONS.

THE OUTER SHEATH COVERING AXONS.

NODES OF RANVIER

MYELINATED

MULTIPLE SCLEROSIS

NERVE

GANGLION

EFFECTORS

NEURAL CIRCUITS OF PATHWAYS

SYNAPSE

REFLEX ACTION (REFLEX ARC)

SENSORY NEURONS

ASSOCIATION NEURON

MOTOR NEURON

PARTS OF THE AXON WHICH HAVE A MYELIN SHEATH.

GAPS IN THE MYELIN SHEATH WHERE THE AXON IS NOT INSULATED.

A COMPLEX OF MANY, PERHAPS THOUSANDS, NEURONS WRAPPED IN CONNECTIVE TISSUE.

A DISEASE OF THE NEURONS IN WHICH PARTS OF THE MYELIN SHEATH ARE DESTROYED AND WHERE DAMAGED SCAR TISSUE DEVELOPS.

MUSCLES OR GLANDS WHICH RESPOND TO A NEURAL IMPULSE.

A GROUPING OF CELL BODIES.

THE JUNCTION BETWEEN THE AXON OF ONE NEURON AND THE DENDRITE OF ANOTHER.

THE SEQUENCE OF NEURONS IN THE NERVOUS SYSTEM.

THE NEURON WHICH RECEIVES THE STIMULUS FROM A RECEPTOR AND SENDS IT DIRECTLY TO AN ASSOCIATION NEURON.

A FIXED AND AUTOMATIC RESPONSE TO A STIMULI.

THE NEURON WHICH RECEIVES THE STIMULUS FROM THE ASSOCIATION NEURON AND SENDS IT TO THE MUSCLE WHICH RESPONDS.

WITHIN THE CENTRAL NERVOUS SYSTEM, RECEIVES THE STIMULUS FROM THE SENSORY NEURON AND SENDS IT TO THE MOTOR NEURON.

RESTING POTENTIAL

SODIUM PUMP

ACTION POTENTIAL

DEPOLARIZATION

WAVE OF DEPOLARIZATION

REPOLARIZATION

THRESHOLD

ALL-OR-NONE

REFRACTORY PERI

PRESYNAPTIC NEURON

POSTSYNAPTIC NEURON

ACETHYLCHOLINE

THE ACTIVE TRANSPORT OF SODIUM OUT OF THE CELL AND THE TRANSPORT OF POTASSIUM IONS INTO THE CELL.

A NEURON WHICH IS READY TO TRANSMIT AN IMPULSE, THE SURFACE OF THE CELL IS NEGATIVELY CHARGED.

THE AREA OF THE CELL SURFACE IN WHICH THE SODIUM IONS ARE MOVING INTO THE CELL THEREBY CHANGING THE POLARITY OF THE CELL SURRACE, SPREADS DOWN THE NEURON IN CHAIN-LIKE FASHION.

A NEURON WHICH RESPONDS TO A STIMULUS, THEREBY REVERSING THE POLARITY OF IONS AT THE SURFACE OF THE CELL FOR A SHORT PERIOD OF TIME.

AS A WAVE OF DEPOLARIZATION PASSES POTASSIUM MOVES OUT OF THE CELL AND SODIUM BACK INTO THE CELL AND SETS UP A RESTING POTENTIAL.

AN ELECTRIC CURRENT WHICH TRAVELS DOWN THE NEURON BY CHANGES IN THE SURFACE POLARITY.

THE STIMULUS IS EITHER ENOUGH TO START AN ACTION POTENTIAL OR IT IS NOT.

THE LEVEL OF INPUT NECESSARY TO TRIGGER A RESTING POTENTIAL TO MOVE INTO AN ACTION POTENTIAL.

A NEURON WHICH ENDS AS A SPECIFIC SYNAPSE.

A VERY SHORT PERIOD OF TIME AFTER AN ACTION POTENTIAL IN WHICH NO NEW IMPULSE CAN BE STARTED.

A NEUROTRANSMITTER (CHEMICAL) WHICH TRANSMITS NEURAL IMPULSES ACROSS SYNAPTIC CLEFTS.

A NEURON WHICH BEGINS AT A SYNAPSE.

CHOLINERGIC NEURONS

CHOLINESTERASE

NOREPINEPHRINE

ADRENERGIC NEURONS

CATECHOLAMINES

NEURAL INTEGRATION

CURCUITS

CONVERGENCE

DIVERGENCE

REVERBERATING CURCUIT

AN ENZYME WHICH BREAKS DOWN ACETHYCHOLINE INTO ITS CHEMICAL COMPONENTS.

CELLS WHICH RELEASE ACETHYLCHOLINE.

NEURONS WHICH RELEASE NORPHINEPHRINE.

A NEUROTRANSMITTER, LEVEL IN THE BRAIN AFFECTS MOOD.

THE BRAIN AND SPINAL CORD ADDING AND SUBTRACTING INCOMING NEURAL SIGNALS TO DETERMINE AN APPROPRIATE RESPONSE.

A CLASS OF COMPOUNDS, SUCH AS NOREPINEPHRINE.

A POSTSYNAPTIC NEURON BEING CONTROLLED BY TWO OR MORE PRESYNAPTIC NEURONS, CONVERGING ON THE FORMER.

THE SPECIFIC PATHWAY OF A NEURAL IMPULSE.

A NEURAL PATHWAY ARRANGED SO THAT A NEURAL COLLATERAL SYNAPSES WITH AN ASSOCIATION NEURON.

A SINGLE PRESYNAPTIC NEURON WHICH STIMULATES MANY POSTSYNAPTIC NEURONS.

CHAPTER 23

RESPONSIVENESS: NERVOUS SYSTEMS

BEFORE READING ACTIVITIES

I. Before you begin to read Chapter 23, review the important points of Chapter 22. Notice that both Chapters 22 and 23 discuss different aspects of the responsiveness of creatures in the animal kingdom.

1. Explain what the cells of the nervous system are and how they work.

2. What are two types of action that cause information to flow through the nervous system? How do these actions work?

3. Explain how impulses are transmitted.

4. Explain neural integration.

5. How are neural circuits organized?

II. Turn to page 471 in the text and read the outline which tells how the chapter about nervous systems is structured. Answer the following questions.

1. Why are there separate sections to discuss invertebrate and human nervous systems? What do you already know about invertebrates that will help you answer this question? (Remember the information in the previous chapter.)

2. What are the sense organs? What do you already know about each of them?

3. What are the major components of the human central nervous system from a biological standpoint. What do you know about each of these from personal experience?

4. What do you think the peripheral nervous system is? How do you expect it to be different from the central nervous system?

III. Vocabulary

1. Review the vocabulary from Chapters 20, 21, and 22. Remember, these chapters are all part of one unit that discusses animal structure and function.

2. Read through the vocabulary terms for this chapter. There are a great many new terms. Use the following hints to help you learn these terms more easily.

   1. Group the words into categories, that is, put terms about the same topic together. For example, all of the terms about the eye should be in one group. all of the terms about the ear should be in another group.

   2. Some of the terms have the same word parts, for example, sympathetic and parasympathetic. How are these words alike and how are they different?

   3. There are two terms that are very similar, afferent and efferent. Be sure you can explain how these are alike and how they are different.

IV. Read the introduction, the summary on pages 491, 492, 493 and 494. Take the Post-Test on pages 494 and 495.

DURING READING ACTIVITIES

I. Use the vocabulary cards as you read.

   1. Circle the terms in you text and underline the definition.
   2. Check to see that you found each term in the text that is on your vocabulary cards.

II. Continue to read the chapters from one major section to the next, marking each section paragraph by paragraph. Be sure to stop at the end of each major section to review what you have read. The example below is from page 486 of the text. Check your own marking to see if you marked the same information. If you did not, try to decide why not.

LEARNING AND MEMORY

Despite extensive research, the mechanisms by which the brain thinks, learns, and remembers are still poorly understood. The human brain differs most markedly from the brains of other animals by the remarkable development of its association areas within the cerebral cortex. Neurons within these areas form highly complex pathways. Damage to association areas can prevent a person from thinking logically, even though he may still be able to hear or even read.

To learn, the brain must be able to (1)focus attention on specific stimuli, whether they be the words on this printed page or an angry wasp buzzing overhead, (2)compare incoming sensory stimuli with stimuli if has encountered before, and (3)store information. Memory is the ability to recall stored information.

III. Be sure to read all of the Figures in the chapter carefully.

   1. When the figures have more than one part, be sure you understand each part.

2. Figures 23-4, 23-5, 23-9, and 23-15 contain information you will probably need to use. You should be able to find and label the parts shown on each of these diagrams and you should be able to explain what each part does.

3. Sometimes it is helpful to create a visual picture or graphic organizer to help you remember information. Look at Table 23-1:

DIVISIONS OF THE HUMAN NERVOUS SYSTEM

I. Central nervous system (CNS)
   A. Brain
   B. Spinal cord
II. Peripheral nervous system (PNS)
   A. Somatic portion
      1. Receptors
      2. Afferent (sensory) nerves - transmit information from receptors to CNS
      3. Efferent nerves - transmit information from CNS to glands and involuntary muscle in organs
   B. Autonomic portion
      1. Receptors
      2. Afferent (sensory) nerves - transmit internal organs to CNS
      3. Efferent nerves - transmit information from CNS to internal organs
         a. Sympathetic nerves - generally stimulate activity that results in mobilization of energy (eg. speeds heartbeat)
         b. Parasympathetic nerves - action results in energy conservation or restoration (eg. slows heartbeat)

What does it tell you? How could you create a "picture" of this information? On the following page is a graphic organizer which shows one way of viewing the information so it will be remembered more easily:

HUMAN NERVOUS SYSTEM

CNS

Brain

Spinal Cord

PNS

Somatic
- Receptors
- Afferent
- Efferent

Autonomic
- Receptors
- Afferent
- Efferent

4. Try to create graphic organizers to "picture" the information in Table 23-2 and 23-3 so you will be able to remember this information more easily.

AFTER READING ACTIVITIES

I. Review your textbook marking.

II. Practice reciting the vocabulary definitions. This is important because if you do not know the vocabulary, you cannot understand or use the information in the text or in lectures.

Answer the questions which follow.

<u>Questions</u>

1. List the trends in bilaterally symmetrical organisms with respect to the nervous system.

2. What are the two main divisions of the human nervous system?

3. What are the components of the central nervous system?

4. What characterizes the peripheral nervous systems?

5. Afferent or sensory nerves transmit information to the ____________ ____________ ____________ from the ____________.

6. The efferent or motor nerves transmit information from ____________ to glands and ____________ ____________ in organs.

7. The efferent nerves of the autonomic system are the ____________ nerves.

8. What are the sense organs in the human body?

9. Which of your senses brings you the most information about your environment?

10. In your eye the ____________ cells are sensitive to color while the ____________ cells are sensitive to light.

11. What four tastes can you distinguish?

12. The ____________________ fluid protects the brain against mechanical injury.

13. What are the two regions distinguished in a cross-section of the spinal cord?

14. What are the three main areas of the cerebral cortex?

15. Match the following:

A. Sensory area	1. Controls voluntary movement
B. Motor area	2. Personality
C. Association area	3. Language ability
	4. Memory
	5. Receives incoming signals from sense organs.
	6. Intelligence and learning

16. Consciousness and alertness are controlled by the __________ _______ ____________.

17. What happens during sleep?

18. The __________ __________ is concerned with the emotional aspects of behavior as well as sexual behavior.

19. Acupuncture may actually stimulate the brain to release __________.

20. List the three brain activities involved in learning.

21. What region of the brain has been identified as the "memory bank"?

22. Environment affects the ability to learn. Parents who provide environmental ___________ for their infants may increase the child's social and academic success.

23. The twelve pairs of ____________ _____________ transmit information dealing with the senses of ___________, ____________, ____________, and ____________.

24. The _______________ roots of the spinal nerves transmit information from the sensory receptors to the ____________ ____________.

25. The autonomic system controls the ____________ ____________. It works ____________.

26. The efferent portion of the autonomic nervous system is further divided into the _______________ and _______________ systems.

27. Examine Table 23-3. Name five drugs which can lead to physical dependence.

Chapter 23 - Answers

1. A. Increased number of nerve cells.
   B. Concentration of nerve cells.
   C. Specialization of function.
   D. Increased number of association neurons and more complex synaptic contacts.
   E. Cephalization - formation of a head.
2. CNS - central nervous system and PNS peripheral nervous system.
3. A complex tubular brain which it continuous with a dorsal, tubular spinal cord.
4. The sensory receptors.
5. Central nervous system, receptors
6. CNS, glands, involuntary muscles
7. Sympathetic, parasympathetic
8. Eyes, ears, nose, taste buds, and touch receptors.
9. Eyes.
10. Cone, rod
11. Sweet, sour, salty and bitter.
12. cerebrospinal fluid
13. White matter- the myelinated axons, and the gray matter - the nerve cell bodies.
14. Sensory, motor, and association areas.
15. A - 5, B - 1, C -2,3,6
16. Reticular activating system.
17. The RAS slows the process of relaying information to the cerebrum.
18. limbic system
19. endorphins
20. A. Focus attention on specific stimuli
    B. Compare incoming stimuli to others encountered previously
    C. Store information
21. No region of the brain has been identified as the memory bank.
22. Stimulation
23. Cranial nerves, smell, sight, hearing, tastes.
24. Dorsal, spinal cord
25. Internal environment, automatically
26. Sympathetic, parasympathetic
27. A. Barbituates
    B. Quaaludes
    C. Meprobamate
    D. Valium
    E. Alcohol
    F. Morphine, heroin

NERVE NET

AFFERENT NEURONS

EFFERENT NEURONS

CEREBRAL GANGLIA

LADDER-TYPE NERVOUS SYSTEM

CIRCUMESOPHAGEAL RING

CENTRAL NERVOUS SYSTEM (CNS)

PERIPHERAL NERVOUS SYSTEM (PNS)

SOMATIC PORTION OF PNS

AUTONOMIC PORTION OF PNS

SENSORY (AFFERENT) NERVES

MOTOR (EFFERENT) NERVES

NERVE CELLS WHICH CONDUCT IMPULSES TOWARD A CENTRAL NERVOUS SYSTEM.

SIMPLIEST ORGANIZED NERVOUS TISSUE. FOUND IN CNIDARIANS. NERVE CELLS SCATTERED THROUGHOUT THE BODY. SENSORY CELLS AND GANGLION CELLS.

CONCENTRATION OF NERVE CELLS IN THE HEAD REGION. PRIMITIVE BRAIN. SEEN IN PLANARIANS.

NERVE CELLS WHICH TRANSMIT IMPULSES AWAY FROM THE CENTRAL NERVOUS SYSTEM AND TO THE EFFECTOR CELLS.

MASS OF 168 MILLION NERVE CELL BODIES AS SEEN IN CEPHALOPODS (e.g., OCTOPUS).

TRANSVERSE NERVES WHICH CONNECT THE BRAIN WITH THE EYESPOTS AND ANTERIOR END OF THE BODY.

THE SENSORY RECEPTORS (TOUCH, AUDITORY, AND VISUAL RECEPTORS). AND THE NERVES WHICH ARE THE COMMUNICATION LINES. LINKS THE CNS WITH THE MUSCLES AND GLANDS. KEEPS BODY IN BALANCE WITH THE EXTERNAL ENVIRONMENT.

CONSISTS OF A COMPLEX TUBULAR BRAIN WHICH IS CONTINUOUS WITH A DORSAL, TUBULAR SPINAL CORD.

REGULATORS OF THE INTERNAL ENVIRONMENT. HAS BOTH SENSORY AND MOTOR NERVES. SYMPATHETIC AND PARASYMPATHETIC NERVES. HELPS MAINTAIN HOMEOSTASIS. WORKS AUTOMATICALLY.

RECEPTORS AND NERVES CONCERNED WITH CHANGES IN THE EXTERNAL ENVIRONMENT. HAS SENSORY AND MOTOR NERVES.

TRANSMIT INFORMATION BACK FROM THE NERVOUS SYSTEM TO THE STRUCTURES THAT MUST RESPOND. ADJUSTS POSITIONS OF THE SKELETEL MUSCLES TO MAINTAIN THE BODY'S WELL-BEING.

TRANSMIT MESSAGES FROM THE RECEPTORS TO THE CNS. INCLUDES RECEPTORS THAT REACT TO CHANGES IN THE EXTERNAL ENVIRONMENT.

SYMPATHETIC NERVES

PARASYMPATHETIC NERVES

SENSE ORGANS

PACINIAN CORPUSCLES

EYE

IRIS

PUPIL

LENS

RETINA

CORNEA

SCLERA

ROD CELLS

ACTION RESULTS IN ENERGY CONSERVATION OR RESTORATION.

PART OF THE AUTONOMIC NERVOUS SYSTEM. GENERALLY STIMULATE ACTIVITY THAT RESULTS IN MOBILIZATION OF ENERGY.

TINY TOUCH RECEPTORS LOCATED IN THE SKIN.

LINK ORGANISMS WITH THE OUTSIDE WORLD AND ENABLES THEM TO RECEIVE INFORMATION ABOUT THE EXTERNAL AND INTERNAL ENVIRONMENT. INCLUDES EYES, EARS, NOSE AND TASTE BUDS. ALSO INCLUDES TOUCH.

COLORED PORTION OF EYE THAT ACTS AS DIAPHRAGM TO REGULATE LIGHT.

FORMS IMAGE.

PART OF EYE WHICH FOCUSES IMAGE.

OPENING OF EYE THROUGH WHICH LIGHT PASSES.

CLEAR COVERING OF IRIS AND PUPIL.

PORTION OF THE EYE MADE UP OF THE LIGHT RECEPTORS.

ONE TYPE OF EYE LIGHT RECEPTOR CELLS WHICH ARE SENSITIVE TO LIGHT BUT NOT TO COLOR. CONTAIN RHODOPSIN.

CLEAR COVERING AROUND EYE, EXCLUDING THE CORNEA.

CONE CELLS

TYMPANIC MEMBRANE (EAR DRUM)

MALLEUS, INCUS, STAPES

SEMICIRCULAR CANALS AND VESTIBULE

TASTEBUDS

OLFACTORY EPITHELIUM

MENINGES

CEREBROSPINAL FLUID

SPINAL CORD

GRAY MATTER

WHITE MATTER

NERVE TRACT

SOUND WAVES CAUSE THIS EAR MEMBRANE TO VIBRATE.

TYPE OF EYE LIGHT RECEPTOR. SENSITIVE TO COLOR. THREE TYPES OF CONE CELLS, EACH SENSITIVE TO A DIFFERENT COLOR--RED, BLUE OR GREEN.

CONCERNED WITH DYNAMIC AND STATIC BALANCE.

THREE SMALL BONES OF MIDDLE EAR WHICH TRANSMIT VIBRATIONS FROM EAR DRUM TO COCHLEA (INNER EAR).

EPITHELIUM OF NOSE INVOLVED IN THE SENSE OF SMELL. CONTAINS SPECIAL OLFACTORY NEURON RECEPTORS. REACT TO AS MANY AS 50 ODORS.

RECEPTORS OF TASTE ON TONGUE AND ON ROOF OF MOUTH. FOUR TASTES ARE DISTINGUISHED: SWEET, SOUR, SALTY AND BITTER.

SPECIAL SHOCK ABSORBING FLUID THAT PROTECTS THE BRAIN AND SPINAL CORD AGAINST MECHANICAL INJURY.

THREE LAYERS OF CONNECTIVE TISSUE WHICH COVER THE BRAIN.

PORTION OF THE SPINAL CORD WHICH CONSISTS OF NERVE CELL BODIES. NEXT TO CAVITY OF THE SPINAL CORD.

HOLLOW CYLINDER OF NERVOUS TISSUE EXTENDING FROM THE BASE OF THE BRAIN TO THE WAIST. FUNCTIONS TO CONTROL REFLEX ACTIVITIES AND TRANSMIT MESSAGES BACK AND FORTH TO THE BRAIN.

BUNDLE OF AXONS.

COVERS THE GRAY MATTER AND CONSISTS OF MYELINATED AXONS OR SPINAL CORD NERVE TRACTS.

BRAIN

CEREBRUM

CEREBRAL CORTEX

SENSORY AREA OF CEREBRAL CORTEX

MOTOR AREAS OF CEREBRAL CORTEX

ASSOCIATION AREAS OF CEREBRAL CORETX

BRAIN WAVES

ALPHA WAVE

BETA WAVES

DELTA WAVES

THETA WAVES

BRAIN STEM

LARGEST, MOST PROMINENT PART OF BRAIN. MORE THAN 70 PERCENT OF BRAIN CELLS. CENTER OF INTELLECT, MEMORY, CONSCIOUSNESS AND LANGUAGE. CONTROLS SENSATION AND MOTOR FUNCTIONS.

MOST COMPLEX ORGAN, WEIGHS 1-4 KILOGRAMS AND CONSISTS OF 25 BILLION CELLS. REQUIRES INPUT OF OXYGEN AND GLUCOSE. REQUIRES 20 PERCENT OF BODY's OXYGEN.

RECEIVES INCOMING SIGNALS FROM SENSE ORGANS.

OUTER PORTION OF CEREBRUM. HIGHLY CONVOLUTED. DIVIDED INTO THREE AREAS: THE SENSORY AREA, THE MOTOR AREA, AND THE ASSOCIATION AREA.

LINK THE SENSORY AND MOTOR AREAS. THEY ARE RESPONSIBLE FOR THOUGHT, LEARNING, INTELLIGENCE, LANGUAGE, MEMORY, JUDGMENT AND PERSONALITY.

CONTROL VOLUNTARY MOVEMENT.

BRAIN WAVE PATTERN ASSOCIATED WITH RESTING AND RELAXED ACTIVITY.

CONTINUOUS ELECTRICAL ACTIVITY OF THE BRAIN. MEASURED IN EEG, ELECTROENCEPHALOGRAM. PRODUCED MAINLY BY THE CEREBRAL CORTEX.

BRAIN WAVES ASSOCIATED WITH SLEEPING.

BRAIN WAVES CHARACTERISTIC OF HEIGHTENED MENTAL ACTIVITY.

CONTAINS VITAL CENTERS THAT CONTROL HEARTBEAT, RESPIRATION, BLOOD PRESSURE, AUDITORY AND VISUAL REFLEXES. CONSISTS OF MEDULLA, PONS, MIDBRAIN, THALAMUS AND HYPOTHALAMUS.

BRAIN WAVES RECORDED MOSTLY IN CHILDREN AND IN SOME ADULTS WHEN UNDER STRESS.

CEREBELLUM

RETICULAR ACTIVATRY SYSTEM (RAS)

LIMBIC SYSTEM

REM SLEEP

ENDORPHINS

LEARNING

MEMORY

CRANIAL NERVES

SPINAL NERVES

DORSAL ROOT OF SPINAL NERVE

VENTRAL ROOT OF SPINAL NERVE

SPINAL GANGLION

COMPLEX NEURONAL PATHWAY IN THE BRAIN STEM. RECEIVES MESSAGES FROM SPINAL CORD NEURONS AND OTHER NEURONS AND COMMUNICATES WITH CEREBRAL CORTEX. MAINTAINS CONSCIOUSNESS AND ALERTNESS.

SECOND LARGEST DIVISION OF BRAIN. REFLEX CENTER FOR MUSCULAR COORDINATION.

RAPID EYE MOVEMENT SLEEP. PERSON DREAMS, ELECTRICAL ACTIVITY INCREASES, CLOSED EYES MOVE RAPIDLY.

ACTION SYSTEM OF BRAIN. CONSISTS OF CEREBRUM, THALAMUS, AND THE HYPOTHALAMUS. AFFECTS EMOTIONAL ASPECTS OF BEHAVIOR, SEXUAL BEHAVIOR, BIOLOGICAL RHYTHMS, AUTONOMIC RESPONSES, AND MOTIVATION.

BRAIN MUST FOCUS ATTENTION ON SPECIFIC STIMULI, COMPARE INCOMING STIMULI TO ONES ENCOUNTERED BEFORE, AND STORE INFORMATION.

ENDOGENOUS MORPHINE-LIKE SUBSTANCES RELEASED BY THE BRAIN.

TWELVE PAIRS OF NERVES WHICH EMERGE FROM THE BRAIN. THESE NERVES TRANSMIT INFORMATION CONCERNED WITH THE SENSES OF SMELL, SIGHT, HEARING AND TASTE. CONTROL VOLUNTARY MUSCLES THAT CONTROL MOVEMENTS OF FACE, MOUTH, TONGUE, PHARYNX AND LARYNX.

THE ABILITY TO RECALL STORED INFORMATION.

CONSISTS OF SENSORY (AFFERENT) FIBERS WHICH TRANSMIT INFORMATION FROM THE SENSORY RECEPTORS TO THE SPINAL CORD.

THIRTY-ONE PAIRS OF NERVES EMERGE FROM THE SPINAL CORD. NAMED FOR THE GENERAL REGION OF THE VERTEBRAL COLUMN FROM WHICH THEY EMERGE: CERVICAL THORACIC, LUMBAR, SACRAL AND COCCYGEAL.

CONSISTS OF THE CELL BODIES OF THE SENSORY NEURONS.

CONSISTS OF THE MOTOR (EFFERENT) FIBERS LEAVING THE CORD LEADING TO THE MUSCLES AND GLANDS.

PSYCHOLOGICAL DEPENDENCE

EUPHORIA

TOLERANCE

ADDICTION

DRUG-INDUCED STATE OF WELL BEING.

USER BECOMES EMOTIONALLY DEPENDENT ON A DRUG.

PHYSIOLOGICAL CHANGES OCCUR IN RESPONSE TO SOME DRUG. ADDICT SUFFERS PHYSICAL ILLNESS WHEN DRUGS ARE WITHDRAWN.

INCREASINGLY LARGER DOSES OF DRUG IS REQUIRED TO GET DESIRED EFFECT.

CHAPTER 24

INTERNAL TRANSPORT

BEFORE READING ACTIVITIES

I. Before you begin to read, review the important points of Chapter 23. If you can, go over these points with another student in the class.

   1. What is the difference between nerve nets and radial systems and bilateral nervous systems?

   2. What is the process of reception? What are the types of receptors in the human body?

   3. What are the components of the human central nervous system? How does each part work: Use your graphic organizer to help you remember.

   4. What are the parts of the peripheral nervous system? How do they work? Be sure to use the information in your graphic organizer.

   5. What are some drugs you should know about? What are the values or concerns associated with these drugs?

II. Turn to page 496 and read the outline about internal transport. Did you notice that Chapter 23 and 24 have sections about both invertebrate and vertebrate animals? Use the outline to answer the following questions about internal transport.

   1. How are vertebrate and invertebrate circulatory systems. different?

   2. What are the components of the blood? What do you already know about plasma, red blood cells, white blood cells, platelets?

   3. What are the blood vessels? What do you think the various types of blood vessels do?

   4. What do you already know about pulse and blood pressure? Is yours normal, high, or low? What difference does it make?

   5. What do you know about the heart?

   6. What is the lymphatic system?

III. Vocabulary

   1. Review the vocabulary from chapter 23. Remember that these terms describe nervous systems.

   2. Read through the vocabulary terms for this chapter. Remember that these terms are about to internal transport.

   3. Group the terms into units that make sense: you might put terms about the heart in one group, terms about blood circulation in another group, terms about the lymphatic system in another group.

4. There are three terms which contain the word part "-phils"; how are these terms alike? How are they different?

5. There are several terms which have the word part "hemo-" in them? What does it mean? How are these terms different?

6. What do the terms with the word part "vasco-" mean?

IV. Read the introductory paragraph on page 497. Read the summary section on pages 519 and 520. Take the Post-Test on page 520.

V. Flip through the chapter by looking at each page briefly.

1. How do the BOLD FACE headings relate to the outline (for example, Plasma on page 499)?

2. Note that there are 18 figures in this chapter. They are an important part of the information discussed in this chapter. be sure you read and understand each of these figures.

DURING READING ACTIVITIES

I. Read the chapter from one BOLDFACE heading to the next BOLDFACE heading. As you complete a section, stop and sum up what the whole section was about. For example, after you read the section, "The Vertebrate Circulatory System" on pages 498 and 499, you should be able to list the vital functions this circulatory system performs.

II. Circle the new vocabulary terms in our text; most of the terms new to you are in darker type, but some may not be. Be sure you circle any term you do not already know and underline the definition. If there is not a vocabulary card for a term you do not know, make one.

III. Be sure that you read at least a whole paragraph before you make any marks on the page. Use the margins to make notes summarizing the important points in your own words.

IV. Reading Figures

1. Look at Figure 24-2 on page 497. What is the difference between the open circulatory system and the closed circulatory system? Find the differences and explain them.

2. Many of the figures in this chapter are summaries of the verbal information contained in the body of the text. It will help you understand the verbal explanation if you read these figures before you read the text.

3. After you read the following figures you should be able to perform the activity listed below.

A. Figure 24-3; trace the blood flow through the system and be able to find and label the various parts of the system.

B. Figure 24-5; Be able to see the different types of blood cells.

C. Figure 24-7; Be able to distinguish between the types of blood vessels.

D. Figure 24-10; Know what the various parts of the heart are and what each part does.

E. Figure 24-14 and 24-15; Be able to show the pattern of blood circulation.

F. Figure 24-16; Be able to locate the parts of the lymphatic system.

AFTER READING ACTIVITIES

I. Check your textbook markings. Did you make notes on each of the figures? Can you use the pictures to explain the various systems? If you did not mark enough information to understand during your review, correct your marking now.

II. Review the vocabulary. Try to relate each of the terms to your own body. Think about what each part of the body does for you, personally.

III. Answer the following questions.

Questions

1. List the three components of a circulatory system.

2. List the components of the vertebrate circulatory system.

3. List the major functions of the vertebrate circulatory system.

4. List the components of whole blood.

5. Match the following plasma components with their corresponding function:

A. Fibrinogen	1. Contains antibodies
B. Serum	2. Plasma proteins which transport hormones and fatty acids.
C. Globulins	3. Blood clotting protein.
D. Albumin	4. Blood minus all red and white blood cells.
	5. Blood minus all red and white blood cells and the blood clotting proteins.

6. Describe a mature red blood cell.

7. Anemia is a condition characterized by a ______________ deficiency. This can be caused by blood loss, decreased red blood cell production, and an increased rate of ______________________________.

8. White blood cells ________________ the body against invading micro- organisms. Also, these cells can pass through the ________________ ____________.

9. Examine Figure 24-5 and answer the following questions.

   A. What is the largest agranular leukocyte? _________________
   B. Name the three granular leukocytes

   1. _____________ 3. ______________
   2. _____________

10. What is the normal ratio of red blood cells to white blood cells?

11. Under what abnormal conditions will the white blood cell population increase?

12. What are the three types of blood vessels?

   1. _______________
   2. _______________
   3. _______________

13. ________________ are the blood vessels through which gas, nutrient and waste exchange occurs, and these vessels are so numerous that virtually every cell in the body is in contact with one.

14. _______________ return blood to the heart while _____________ take blood away from the heart.

15. Vasoconstriction and _________________ help maintain the appropriate blood ______________.

16. Normally, most of the blood supply is directed to the ___________, _______________, and brain.

17. _______________ is the phase of the heart cycle characterized by relaxation of the ventricles.

18. The pulse indicates the number of _______________ _______________ per minute.

19. Examine Figure 24-12. Match the following:

A. P Wave	1. Closing of the atrial valve
B. QRS Complex	2. Recovery of the ventricle
C. T Wave	3. Spread of impulse to the ventricles prior to contraction
	4. Spread of impulse through atria prior to atrial contraction

20. Examine Figure 24-13. Name three situations which could decrease blood pressure.

21. Why do physicians recommend low-salt diets for patients who have heart conditions?

22. What is the most common cardiovascular problem in the U.S.A.?

23. The _____________ circulation connects the heart with all the tissues of the body.

24. The _____________ artery is the only artery that carries deoxgenated blood.

25. What is the sequence of structures through which the blood flows during the pulmonary circulation? Start with the right atrium.

26. _______________ disease is the number on cause of death in the United States and other industrial societies?

27. List the high risk factors for heart disease.

Chapter 24 - Answers

1. Blood, heart, blood vessels
2. Ventral heart, blood vessels, blood lymph, lymph vessels, thymus, spleen, and liver
3. Transports nutrients, transports oxygen, transports metabolic wastes, transports hormones, maintains fluid balance, transports elements of immumne system, and distributes metabolic heat.
4. Plasma, red blood cells, white blood cells, and platelets.
5. A. 3
   B. 5
   C. 1
   D. 2
6. Very small cell filled with hemoglobin. Surface maximized for gas exchange. Mammalian red blood cells lack a nucleus and have a short life span of 120 days. 2.4×10^6 red blood cells destroyed per second.
7. Hemoglobin (red blood cell), red blood cells destruction
8. Defend, capillary walls
9. A. Monocyte
   B. Neutrofils, basophils, and eosinophils
10. 700 to 1
11. Leukemia and certain bacterial infections
12. Arteries, capillaries, veins
13. Capillaries
14. Veins, arteries
15. Vasodilation, pressure
16. Liver, kidneys
17. Diastole
18. heart beats
19. A. 4
    B. 3
    C. 2
20. Decreased blood volume (hemorrhage), decreased cardiac output, and decreased peripheral resistance
21. Because increased salt causes water to enter the blood, increasing the blood volume and therefore the blood pressure
22. Hypertension
23. Systemic
24. Pulmonary
25. Right atrium, right ventricle, pulmonary artery, pulmonary capillaries, pulmonary vein, and left atrium
26. Cardiovascular
27. High blood cholesterol levels, hypertension, cigarette smoking, and diabetes mellitus

CIRCULATORY SYSTEM

OPEN CIRCULATORY SYSTEM

HEMOCOEL

CLOSED CIRCULATORY SYSTEM

BLOOD

PLASMA

ALBUMIN PROTEINS

GLOBULINS

FIBRINOGEN

SERUM

RED BLOOD CELLS (ERYTHROCYTES)

RED BONE MARROW

HEART PUMPS BLOOD INTO VESSELS THAT HAVE OPEN ENDS. ARTHROPODS AND MOST MOLLUSKS. NOT AS EFFICIENT AS A CLOSED SYSTEM.

COMPLEX ANIMALS HAVE THIS SYSTEM FOR INTERNAL TRANSPORT. CONSISTS OF BLOOD FLUID CONNECTIVE TISSUE AND CELLS DISPERSED IN FLUID, A PUMPING DEVICE - HEART, AND A SYSTEM OF BLOOD VESSELS THROUGH WHICH THE BLOOD CIRCULATES.

BLOOD FLOWS THROUGH A CONTINUOUS SYSTEM OF BLOOD VESSELS.

BLOOD CAVITY INTO WHICH BLOOD IS DEPOSITED WHICH BATHES THE CELLS OF THE BODY.

COMPONENT OF WHOLE BLOOD, THE FLUID IN WHICH THE BLOOD CELLS ARE SUSPENDED. CONSISTS OF 92 PERCENT WATER, PLASMA PROTEINS, SALTS, TRANSPORTED NUTRIENTS, WASTES, HORMONES AND DISSOLVED GASES.

CONSISTS OF PLASMA RED BLOOD CELLS, WHITE BLOOD CELLS AND PLATLETS.

PLASMA PROTEINS MANY OF WHICH ARE ANTIBODIES TO PROVIDE IMMUNITY AGAINST INVADING DISEASE ORGANISMS.

PLASMA PROTEINS. TRANSPORT HORMONES, FATTY ACIDS AND OTHER COMPOUNDS.

PLASMA FROM WHICH ALL THE BLOOD CLOTTING PROTEINS HAVE BEEN REMOVED.

ONE OF THE MAJOR PLASMA PROTEINS, INVOLVED IN THE CLOTTING PROCESS.

SITE OF PRODUCTION OF RED BLOOD CELLS.

RED CELLS SUSPENDED IN BLOOD. CONTAINS RED PIGMENT HEMOGLOBIN WHICH TRANSPORTS OXYGEN.

HEMOGLOBIN

OXYHEMOGLOBIN

ANEMIA

WHITE BLOOD CELLS
(LEUKOCYTES)

AGRANULAR LEUKOCYTES

MONOCYTES

MACROPHAGES

GRANULAR LEUKOCYTES

NEUTROPHILS

EOSINOPHILS

BASOPHILS

LEUKEMIA

FULLY OXYGENATED HEMOGLOBIN. EACH OF THE FOUR IRON ATOMS HAS A MOLECULE OF OXYGEN.

RED PIGMENT IN RED BLOOD CELLS: TRANSPORTS OXYGEN.

CELLS WHICH DEFEND THE BODY AGAINST INVADING BACTERIA AND OTHER FOREIGN SUBSTANCES. THESE CELLS ARE PASSED THROUGH THE CAPILLARY WALLS AND WANDER IN TISSUES. FORMED IN RED BONE MARROW.

CONDITION CHARACTERIZED BY REDUCED NUMBER OF RED BLOOD CELLS, THEREFORE LESS HEMOGLOBIN. THREE GENERAL CAUSES OF ANEMIA ARE LOSS OF BLOOD, DECREASED PRODUCTION OF RED BLOOD CELLS AND INCREASED RATE OF RED BLOOD CELL DESTRUCTION.

IMMATURE CELLS WHICH DEVELOP IN TISSUES. INCREASE SIZE AND BECOME MACROPHAGES.

LACK GRANULES. MONOCYTES AND LYMPHOCYTES.

ANOTHER GROUP OF WHITE BLOOD CELLS WHICH INCLUDES NEUTROPHILS, EOSINOPHILS, AND BASOPHILS.

GIANT SCAVENGER CELLS OF THE BODY WHICH DESTROY BACTERIA AND DEAD CELLS.

LEUKOCYTES INVOLVED IN ALLERGIC REACTIONS.

LEUKOCYTES WHICH SEEK OUT AND DESTROY BACTERIA.

FORM OF CANCER IN WHICH WHITE BLOOD CELLS MULTIPLY IN RED BONE MARROW AND MANY OF THESE CELLS DO NOT MATURE.

LEUKOCYTES WHICH CONTAIN LARGE AMOUNTS OF HISTAMINE WHICH IS RELEASED IN INFECTED TISSUES AND ALLERGIC REACTIONS. CONTAIN HEPARIN WHICH MAY PREVENT ERRONEOUS CLOTTING.

PLATELETS

ARTERY

ARTERIOLES

CAPILLARIES

VEINS

VASOCONSTRUCTION

VASODILATION

HEART

ATRIUM, ATRIA (PL)

VENTRICLE

PERICARDIUM

VALVES

BLOOD VESSEL WHICH CARRIES BLOOD AWAY FROM THE HEART AND TOWARDS OTHER TISSUES. THICK WALLS.

SMALL FRAGMENTS OF CYTOPLASM THAT SEPARATE FROM LARGE CELLS IN THE BONE MARROW. THESE STRUCTURES ARE INVOLVED IN THE BLOOD CLOTTING PROCESS.

SMALLEST BLOOD VESSELS WHICH CARRY BLOOD IN ORGANS. THIN WALLS. EXCHANGE OF MATERIALS THROUGH THESE WALLS.

SMALLER BRANCHES OF ARTERIES WHICH ARE FOUND IN ORGANS. THESE VESSELS DELIVER BLOOD INTO SMALLER CAPILLARIES.

CONSTRUCTION OF ARTERIOLE BY SMOOTH MUSCLE CONTRACTION.

BLOOD VESSELS WHICH TRANS-PORT BLOOD BACK TO THE HEART. THICK WALLS.

BLOOD PUMP. CAN PUMP 5 TO 35 LITERS OF BLOOD PER MINUTE. CARDIAC MUSCLE, HOLLOW AND CONSISTS OF FOUR CHAMBERS.

RELAXATION OF SMOOTH MUSCLES OF ARTERIOLES.

PART OF THE HEART WHICH PUMPS BLOOD OUT INTO SPECIFIC BLOOD VESSELS.

PART OF THE HEART WHICH RECEIVES BLOOD. HUMAN HEART HAS TWO ATRIA.

STRUCTURES IN THE HEART AND LARGER ARTERIES WHICH PREVENT BACKFLOW OF BLOOD.

MEMBRANE WHICH FORMS BOUNDRY OF CAVITY WHICH SURROUNDS THE HEART.

PACEMAKER (SA OR SINOATRICAL NODE)

HEARTBEAT

INTERCALATED DISCS

CARDIAC OUTPUT

HEARTH RATE

SYSTOLE

DIASTOLE

ELECTROCARDIOGRAPH

PULSE

BLOOD PRESSURE

PERIPHERAL RESISTANCE

HYPERTENSION

SERIES OF CONTRACTIONS OF THE DIFFERENT HEART COMPONENTS STARTING WITH THE PACEMAKER. NEXT THE ATRIA CONTRACTS AND THEN VENTRICLES.

POINT WHERE EACH HEARTBEAT BEGINS.

THE VOLUME OF BLOOD PUMPED BY ONE VENTRICLE. NORMAL OUTPUT IS 5 LITERS PER MINUTE.

DENSE BANDS WHICH SEPARATE THE CARDIAC MUSCLE CELLS. SPECIALIZED TIGHT JUNCTION BETWEEN CELLS.

THE PHASE OF THE HEART'S CYCLE WHEN THE VENTRICLES CONTRACT.

NUMBER OF TIMES HEART CONTRACTS PER MINUTE. NORMAL FOR HUMANS IS ABOUT 70 BEATS PER MINUTE.

MACHINE USED TO AMPLIFY THE ELECTRICAL SIGNALS GENERATED DURING HEART CONTRACTION. PRODUCES A RECORD CALLED THE ELECTROCARDIOGRAM EKG OR ECG).

PHASE OF THE HEART CYCLE WHEN SEMILUNAR VALVES CLOSE AND THE VENTRICLES RELAX.

FORCE EXERTED AGAINST THE INNER WALLS OF THE BLOOD VESSELS.

THE EXPANSION AND RETRACTION OF THE ELASTIC ARTERY WALL AS BLOOD PASSES THROUGH IT. NUMBER OF PULSATIONS INDICATES THE NUMBER OF HEARTBEATS.

HIGH BLOOD PRESSURE. WHEN DIASTOLIC PRESSURE INCREASES TO CONSISTENT READING OVER 95mm MERCURY.

THE RESISTANCE TO BLOOD FLOW CAUSED BY BLOOD VISCOSITY, AND THE FRICTION BETWEEN THE BLOOD AND THE WALL OF THE BLOOD VESSEL.

VEIN VALVES

BARORECEPTORS

ANGIOTENSINS

RENINS

PULMONARY CIRCULATION

SYSTEMIC CIRCULATION

AORTA

ATHEROSCLEROSIS

ISCHEMIC TISSUE

ISCHEMIC HEART DISEASE

MYOCARDIAL INFARCTION

CORONARY OCCLUSION

TINY PRESSURE RECEPTORS IN SOME ARTERIES WHICH ARE SENSITIVE TO CHANGES IN BLOOD PRESSURE.

VALVES WHICH PREVENT THE BACKFLOW OF BLOOD.

HORMONE RELEASED FROM KIDNEYS WHEN BLOOD PRESSURE IS LOW.

HORMONES WHICH REGULATE BLOOD PRESSURE WHICH ARE VASO-CONSTRICTORS.

BLOOD VESSELS WHICH CONNECT THE HEART WITH ALL THE TISSUES OF THE BODY.

BLOOD VESSELS WHICH CONNECT THE HEART AND LUNGS.

HARDENING OF THE ARTERIES AS A RESULT OF LIPID DEPOSITION.

THE LARGEST ARTERY OF THE BODY.

CORONARY ARTERY NARROWS AND IS NOT ABLE TO DELIVER SUFFICIENT OXYGEN TO HEART DURING INCREASED NEED.

A TISSUE LACKING AN ADEQUATE BLOOD SUPPLY WHICH RESULTS IN TOO LITTLE OXYGEN REACHING THE TISSUE.

A CLOT IN THE CORONARY ARTERY. CAN LEAD TO CARDIAC ARREST IF VERY LARGE.

SERIOUS (OFTEN FATAL) FORM OF ISCHEMIC HEART DISEASE. PORTION OF CARDIAC MUSCLE DEPRIVED OF OXYGEN. SYNONYMOUS WITH HEART ATTACK.

CORONARY ARTERIES

CARDIOPULMONARY RESUSCITATION

LYMPHATIC SYSTEM

LYMPH

LYMPH NODES

INTERSTITIAL FLUID

EDEMA

METHOD FOR AIDING VICTIMS OF ACCIDENTS OR HEART ATTACKS WHO HAVE SUFFERED CARDIAC ARREST AND RESPIRATORY ARREST.

SUPPLY BLOOD TO THE HEART MUSCLE ITSELF. BRANCH OFF THE AORTA.

EXCESSIVE TISSUE FLUID.

ACCESSORY TO THE CIRCULATORY SYSTEM. FEEDS INTO BLOOD CIRCULATION. THREE FUNCTIONS ARE TO COLLECT AND RETURN EXCESS TISSUE FLUID TO BLOOD, DEFEND THE BODY AGAINST DISEASE ORGANISMS, AND ABSORB LIPIDS FROM THE DIGESTIVE SYSTEM.

PLASMA WHICH IS FORCED OUT OF THE CAPILLARY WALLS. IT CONTAINS OXYGEN AND NUTRIENTS AND BATHES THE CELLS. RECEIVES WASTE FROM CELLS WHICH EVENTUALLY RETURNS TO THE BLOOD.

SMALL ORGANIZED MASSES OF LYMPH TISSUE WHICH HAVE TWO FUNCTIONS. THEY SLOWLY FILTER THE LYMPH AND PRODUCE LYMPHOCYTES WHICH ARE CONCERNED WITH THE IMMUNE RESPONSE.

ACCUMULATION OF TISSUE FLUID IN AN AREA WHERE LYMPH VESSELS BECOME BLOCKED.

CHAPTER 25

INTERNAL DEFENSE: IMMUNITY

BEFORE READING ACTIVITIES

I. Before you begin to read, review the major concepts presented in Chapter 24. Reviewing information is an important step in remembering because you need to make sure the information you studied is transferred from your short-term to your long term memory because information in short-term memory remains for only a brief time and then is lost. To increase the chances of remembering information (transferring it from short-term to long-term memory) you need to practice using it. Reviewing information is a form of practice. To review the information on internal transport do the following:

1. Turn back to the outline on page 496. Try to recite the main ideas associated with each point in the outline.

2. If you can't remember what information goes with any idea, turn to the appropriate portion of the text and use your textbook marking, that is, your underlining and margin notes, to help you.

3. You may also want to review the figures in the text.

4. Be sure to test yourself on the vocabulary by either saying the definitions out loud or writing them down.

II. Turn to page 522 of your textbook and look over the outline which lists the important concepts to be discussed. Answer the following questions.

1. What do you think "self and nonself" means?

2. Why is there a short section on immunity in invertebrates and a long section about immunity in vertebrates?

3. What do you think is the difference between "specific" and "nonspecific" defense mechanisms? Do you have any defense mechanisms that you are aware of?

4. What is the difference between something that is active and something that is passive? Are you an active or a passive person?

III. Read the Learning Objectives below and answer the questions which follow.

Learning Objectives

After you have studied this chapter you should be able to:

1. Compare in general terms the types of immune responses in invertebrates and vertebrates.
2. Distinguish between specific and nonspecific immune responses.
3. Describe nonspecific defense mechanisms, such as inflammation and phagocytosis, and summarize their role in the defense of the body.

4. Contrast T and B lymphocytes with respect to life cycle and function.

5. Describe the mechanisms of cell-mediated immunity, including development of memory cells.
6. Define the terms antigen and antibody and describe how antigens stimulate immune responses.

1. In what part of the chapter will you be most likely to find the information that will enable you to compare immune reponses in invertebrates and vertebrates? Page______.
2. In what part of the chapter will you learn about specific and nonspecific immune responses (Objective 2)? Page______.
3. Read each objective (only the first 6 are given here) and determine where in the chapter you will find the information. Is it given in the verbal portion of the text? In the figures? In the tables?

IV. Read the introductory section on page 523, and the summary on pages 539 and 540. Read through your vocabulary terms in this text. Then take the Post-Test on page 540.

V. Tear out the vocabulary cards in this text and start to organize them into groups that make sense to you. One of the best ways to help yourself remember is to use your own personal organization. Some of the things you should notice as you organize:

1. The word part "-cyte" is used; what does it mean? Where did you already learn it?
2. The word part "lymph-" is used in more than one word. How are these words alike? Different?
3. Be sure to learn the specialized meanings of what seem to be common terms such as "complement" or "immune response".

VI. Flip through the pages of Chapter 25 (523-539).

1. Note the relationship between the BOLD FACE headings and the Dark Face subheadings. These should tell you which are the most important concepts and which are the major ideas within each larger concept.

2. You should also note the relationship between these headings and the outline on page 522.

3. There are 9 figures in this chapter and most of them look different from those in the previous chapters. What are the differences? Most of these are flow charts which show the order in which the immune system operates; in many other chapters there were illustrations of how things looked or the components of something (such as a cell).

DURING READING ACTIVITIES

I. Before you start to read the section "Self and Nonself" think of at least one question you expect to find an answer to when you read the section. Examples might be, "What does self and nonself have to do with immunity?" or "What is the difference between self and nonself?" When you finish reading the section, stop and think about the answer to the question. Did you find an answer? What is it?

II. Follow the same procedure before you start to read the section on "Immunity In Invertebrates". Ask yourself at least one question and then read to see if you can find the answer. Do this before you begin each new section.

III. Use the darker type in the text as a guide to which terminology is important. Be sure to circle those terms and any others you don't know. Underline the definitions and make vocabulary cards for any terms not already included in the vocabulary section.

IV. Be sure to mark the text by first reading a whole paragraph and then underlining the main ideas and, finally, making notes in the margin.

V. Be sure to read the figures carefully. Some of the things to look for include:

1. Notice how the arrows are drawn. What does an arrow pointing in one direction only mean? What does it mean if there are two arrows which begin at the same point but end in different places?

2. How does Figure 25-1 differ from 25-2? Where does Figure 25-2 fit into Figure 25-1? How are these two figures alike?

3. What do the two different types of arrows (solid line and broken line) mean in part A of Figure 25-4? Why are there two separate sets of arrows in this figure?

4. How is Figure 25-7 different from the previous figures? It does not have arrows and it does not show the order in which something progresses. What does it show?

5. Look at Figures 25-5 and 25-6. What does each depict? How are they alike? How are they different? Where do these figures fit into the summary of defense mechanisms given in Figure 25-1? Can you explain how each of these immunity systems works?

AFTER READING ACTIVITIES

I. Review your marking. Is what you marked and wrote in each section enough to remind you what the important facts and ideas in each section are? If not, be sure to correct your marking.

II. Did you categorize your vocabulary terms before you began reading? If so, how many of the terms are in groups that make sense to you now? Do you need to regroup? Continue to study these terms in groups of 5-7 at a time. Remember that you need to practive reciting each group of terms several times each day to be able to remember them.

III. Answer the following questions.

Questions

1. ____________ defense mechanisms involve the body producing a specific antibody to combat a particular pathogen. In contrast, ____________ defense mechanisms defend the body against a variety of pathogens.

2. An ____________________ is a substance which stimulates an immune response in an organism.

3. Examine Figure 25-1 and answer the following questions.

   A. What are the two types of defense mechanisms?

      1. ____________ 2. ____________

   B. What are some of the body's barriers to pathogens?

   C. List three specific nonspecific defense mechanisms.

      1. ____________
      2. ____________
      3. ____________

   D. List two specific defense mechanisms.

      1. ____________
      2. ____________

4. Most invertebrate have only ____________ defense mechanisms.

5. All vertebrates have both ____________ and ____________ defense mechanisms.

6. In humans the ____________ is populated by millions of microorganisms which do not reproduce.

7. Microorganisms which enter with the food are destroyed by the ____________ secretions of the stomach.

8. ____________ are protein molecules which stimulate some cells to produce antiviral substances. These protein molecules are of great interest to drug companies.

9. What causes a fever?

10. What is one of the main functions of inflammation?

11. Are neutrophils immortal? Explain.

12. Do specific immune responses act immediately to destroy invading pathogens?

13. What are the two types of specific immunity?

14. Why is it possible for people to not get certain diseases (measles, chicken pox) twice?

15. Examine Figure 255 and answer the following questions.

A. What stimulates the "small" lymphocytes to become sensitized and mitotically active?

B. What is the fate of the memory cells?

C. What is the fate of the killer T lymphocytes?

D. Killer T lymphocytes release a substance which kills the pathogens, what is this substance?

16. The B lymphocytes are responsible for the ______________ immunity.

17. Examine Figure 25-6, antibody-mediated immunity.

A. Which cells secrete specific antibody?

B. Do all the stimulated B lymphocytes differentiate to form plasma cells?

C. What are the functions of the memory cells?

D. What happens to the antibodies?

18. Examine Figure 25-7 and answer the following questions.

    A. How many light chains on a typical antibody molecule?

    B. What are structural building blocks of the heavy and light chains?

    C. Typically, how many binding sites on an antibody molecule?

19. What are the two most common types of antibodies found in humans?

20. Antigen-antibody complex may ______________ the pathogen and destroy its ability to function.

21. The antigen-antibody complex sometimes stimulates ________ of the pathogen.

22. There is a tremendous ______________ of antibodies.

23. What are the functions of the thymus gland?

24. When a person is injected with some vaccine to prevent some disease such as measles, this is ________________ immunity.

25. When a person receives an injection of gamma globulin from a physician, this is an example of _________ ___________.

26. Both naturally induced and artificially induced _________ immunity cause the production of memory cells and are long-lived. (Hint - see Table 25-1).

27. Do allergens stimulate a response in all individuals? Explain.

28. Examine Figure 25-8 and answer the following questions.

   A. Histamine is released from which type of cells and where are these cells located?

   B. What event starts the entire allergic reaction?

29. Examine Figure 25-9. Which cells are involved in the body's response to a transplanted foreign organ?

Chapter 25 - Answers

1. Specific, nonspecific
2. Antigen or immunogen
3. A. Specific and nonspecific
   B. Skin, mucous lining of respiratory passages
   C. inflammation, phagocytosis, interferon
   D. cell-mediated immunity and antibody-mediated immunity
4. Nonspecific
5. Nonspecific, specific
6. Skin
7. Acidic
8. Interferons
9. When there is a widespread inflammation, neutrophils and macrophages release endogenous pyrogens which affect the brain regulation of body temperature. The elevated temperature interferes with the bacterial metabolism.
10. Increased phagocytosis
11. No. After a neutrophil has ingested 20 bacteria, it dies.
12. No. Several days are required to activate these mechanisms.
13. Cell-mediated immunity and antibody-mediated immunity.
14. The orginal infection stimulated the production of T lymphocytes. Some of these cells remained as memory cells.
15. A. Macrophage carries antigen to the lymphocytes
    B. They remain in the lymph node
    C. They leave the lymph node and migrate to infection site
    D. Lymphokines
16. Antibody-mediated
17. A. Plasma cells,
    B. Some B lymphocytes become memory cells
    C. They continue to secrete antibodies after the infection is gone, and they enable the body to quickly respond to the same antigen if it enters again.
    D. The antibodies combine with the antigens on the surface of the pathogen and this leads to destruction of the pathogen.
18. A. Two
    B. Amino acids
    C. Two
19. Ig G - blood, fight off viruses, bacteria and fungi
    Ig A - principal antibodies in mucous secretions
20. Inactivate
21. Phagocytosis
22. Diversity
23. 1. Confers immunological competence on T lymphocytes
    2. Endocrine gland secretion
24. Induced
25. Passive immunity
26. Active
27. No. Only in those people who are allergic
28. A. Mast cells, nasal passage
    B. The allergens on pollen surfaces cause the release of Ig E from sensitized plasma cells
29. T lymphocytes

PATHOGENS

IMMUNITY OR IMMUNE RESPONSES

IMMUNOLOGY

NONSPECIFIC DEFENSE MECHANISMS

SPECIFIC DEFENSE MECHANISMS

ANTIGEN OR IMMUNOGEN

INTERFERONS

INFLAMMATORY RESPONSE

EDEMA

FEVER

ENDOGENOUS PYROGENS

PHAGOSOME

DEFENSE MECHANISMS AGAINST PATHOGENS. IMMUNITY MEANS SAFE.

DISEASE CAUSING ORGANISMS.

FIGHT OFF A VARIETY OF PATHOGENS. PREVENT ENTRY OF PATHOGENS AND DESTROY THEM.

THE STUDY OF THE DEFENSE SYSTEMS WHICH COMBAT PATHOGENS.

A SUBSTANCE WHICH IS CAPABLE OF STIMULATING AN IMMUNE RESPONSE.

DEFENSE AGAINST A PARTICULAR PATHOGEN WHICH ENTERS THE BODY. ANTIBODIES (SPECIFIC PROTEINS) PRODUCED TO DESTROY PATHOGENS.

PATHOGENS INVADE TISSUE, BLOOD VESSELS DILATE, INCREASED BLOOD FLOW. FLUID ACCUMULATES AND INCREASED PHAGOCYTOSIS.

A GROUP OF PROTEINS WHICH ARE SECRETED BY CELLS WHICH ARE INFECTED BY VIRUSES, BACTERIA, FUNGI OR PROTOZOA. THESE PROTEINS STIMULATE OTHER CELLS TO PRODUCE ANTIVIRAL SUBSTANCES.

CLINICAL SYMPTOM OF WIDE SPREAD INFLAMMATORY RESPONSE. ELEVATED BODY TEMPERATURE.

THE SWELLING WHICH OCCURS AT AN INFECTED SITE. THIS IS DUE TO INCREASED VOLUME OF INTERSTITIAL FLUID. CAUSES PAIN.

IN NEUTROPHILS AND MICROPHAGES, THE MEMBRANE BOUND INTRACELLULAR VESTICLE CONTAINING AN INGESTED BACTERIUM WILL FUSE WITH THE LYSOSOMES. THE LYSOSOMAL ENZYMES WILL DESTROY THE BACTERIUM.

PROTEINS RELEASED BY NEUTROPHILS AND MACROPHAGES WHICH CAUSE THE BODY TEMPERATURE TO INCREASE.

CELL-MEDIATED IMMUNITY

ANTIBODY-MEDIATED IMMUNITY

T-LYMPHOCYTES
(T CELLS)

B LYMPHOCYTES (B CELLS)

COMPETENT LYMPHOCYTE

KILLER T-LYMPHOCYTES

LYMPHOKINES

MEMORY CELLS

PLASMA CELLS

B MEMORY CELLS

IMMUNOGLOBULINS

ANTIGENIC DETERMINANT

LYMPHOCYTES PRODUCE SPECIFIC ANTIBODIES DESIGNED TO DESTROY THE PATHOGEN.

LYMPHOCYTES ATTACK THE INVADING PATHOGEN DIRECTLY.

CELLS RESPONSIBLE FOR ANTIBODY-MEDIATED IMMUNITY. PROBABLY PROCESSED IN THE BONE MARROW. THOUSANDS OF DIFFERENT B LYMPHOCYTES.

CELLS PRODUCED IN THE BONE MARROW (EMBRYONIC LIVER) WHICH MIGRATE TO THE THYMUS GLAND FOR PROCESSING. THESE CELLS THEN BECOME RESPONSIBLE FOR CELL-MEDIATED IMMUNITY AND DESTROY VIRUSES OR FOREIGN CELLS THAT ENTER THE BODY. ACCUMULATE IN LYMPH TISSUE.

CELLS WHICH COMBINE WITH THE ANTIGEN ON SURFACE OF THE INVADING CELL AND KILL IT.

PARTICULAR LYMPHOCYTES WHICH RESPOND TO A PARTICULAR ANTIGEN WHICH IS DELIVERED BY THE ENGULFING MACROPHAGE.

AFTER T LYMPHOCYTES ARE ACTIVATED AND DIVIDE, SOME CELLS REMAIN BEHIND TO ATTACK THE PATHOGEN IF IT SHOULD TRY TO ATTACK AGAIN.

GLYCOPROTEINS RELEASED BY KILLER T LYMPHOCYTES WHICH KILL THE INVADING CELL. ALSO INCREASE THE INFLAMMATORY RESPONSE.

B LYMPHOCYTES WHICH DO NOT FORM PLASMA CELLS BUT CONTINUE TO PRODUCE ANTIBODY AFTER THE INFECTION IS OVERCOME.

ACTIVATED B LYMPHOCYTES DIVIDE AND PRODUCE LARGE NUMBERS OF IMMUNOLOGICALLY IDENTICAL LYMPHOCYTES. MOST OF THESE CELLS INCREASE IN SIZE AND DIFFERENTIATE INTO THE PLASMA CELLS WHICH PRODUCE ANTIBODIES THAT ARE SECRETED.

AMINO ACID REGION OF AN ANTIGEN WITH A PARTICULAR CONFORMATION RECOGNIZED BY AN ANTIBODY OR CELL RECEPTOR.

HIGHLY SPECIFIC PROTEINS WHICH ARE OFTEN PRODUCED IN RESPONSE TO A SPECIFIC ANTI-GEN. EACH ONE CONSISTS OF FOUR POLYPEPTIDES, TWO IDENTI-CAL HEAVY CHAINS AND TWO IDENTICAL LIGHT CHAINS.

HAPTEN

IgG

IgA

COMPLEMENT SYSTEM

OPSONIZATION

THYMUS GLAND

ACTIVE IMMUNITY

IMMUNIZATION

PASSIVE IMMUNITY

HYPERSENSITIVITY

ALLERGENS

ALLERGIC ASTHMA

ONE OF THE FIVE TYPES OF ANTIBODIES. IN HUMANS 75 PERCENT OF THE ANTIBODIES IN THE BLOOD ARE IN THIS GROUP. THEY ARE PART OF THE GAMMA GLOBULIN FRACTION OF THE PLASMA, AND THEY CONTRIBUTE IMMUNITY AGAINST VIRUSES, BACTERIA, AND SOME FUNGI.

SUBSTANCES FOUND IN DUST AND CERTAIN DRUGS WHICH ARE TOO SMALL TO BE ANTIGENIC. THEY BECOME ANTIGENIC BY COMBINING WITH LARGER PROTEIN MOLECULES.

A SERIES OF REACTIONS STIMULATED BY THE ANTIGEN - ANTIBODY COMPLEX WHICH ARE FIXED AND THEN DESTROY THE PATHOGENS.

THESE IMMUNOGLOBULINS ARE THE MAIN ANTIBODIES FOUND IN MUCOUS SECRETIONS. PROTECT AGAINST INHALED AND INGESTED PATHOGENS.

GLAND WHICH HAS TWO FUNCTIONS: THYMUS CONFERS IMMUNOLOGICAL COMPETENCE ON T LYMPHOCYTES, AND THE THYMUS FUNCTIONS AS AN ENDOCRINE GLAND AND SECRETES THYMOSIN.

ANTIGEN-ANTIBODY COMPLEX COATS PATHOGEN AND INCREASES THE LIKELIHOOD THAT PHAGOCYTOSIS WILL OCCUR.

DEVELOPMENT OF IMMUNITY AS A RESULT OF BEING INJECTED WITH A VACCINE. INDUCED IMMUNITY.

IMMUNITY DEVELOPED FROM EXPOSURE TO ANTIGENS.

STATE OF ALTERED IMMUNE RESPONSE THAT IS HARMFUL TO THE BODY. INCLUDES ALLERGIC REACTIONS AND AUTOIMMUNE DISEASE.

AN INDIVIDUAL RECEIVES ANTIBODIES FROM ANOTHER ORGANISM BORROWED IMMUNITY.

AN ALLERGEN-REAGIN REACTION WHICH OCCURS IN THE BRONCHIOLES OF THE LUNGS.

MILD ANTIGENS WHICH STIMULATE THE PRODUCTION OF ANTIBODIES IN ALLERGIC INDIVIDUALS WHO CONSTITUTE 15 PERCENT OF THE POPULATION.

SYSTEMIC ANAPHYLAXIS

ANTIHISTAMINES

DESENSITIZATION THERAPY

AUTOIMMUNE DISEASE

HISTOCOMPATABILITY ANTIGENS

AUTOGRAFT

HOMOGRAFT

GRAFT REJECTION

IMMUNOLOGICALLY PRIVILEGED

DRUGS WHICH BLOCK THE EFFECTS OF HISTAMINE.

DANGEROUS ALLERGIC REACTION, OCCURS QUICKLY AND AFFECTS THE ENTIRE BODY. SHOCK AND DEATH CAN OCCUR.

DISEASES WHICH RESULT FROM FAILURE OF SELF TOLERANCE. INCLUDES RHEUMATOID ARTHRITIS, MYASTHENIA GRAVIS, AND SYSTEMIC LUPUS ERYTHEMATOSIS.

MEANS TO HELP ALLERGIC PEOPLE. VERY SMALL QUANTITIES OF ANTIGENS ARE ADMINISTERED SLOWLY OVER LONG PERIOD OF TIME ANTIBODIES FORM.

TISSUE TRANSPLANTED FROM ONE LOCATION TO ANOTHER IN THE SAME INDIVIDUAL.

GROUPS OF ANTIGENS THAT ARE A PART OF AN INDIVIDUAL'S TISSUES AND ORGANS.

BODY'S IMMUNE REACTION TRANSPLANTED FOREIGN GRAFT.

GRAFT MADE BETWEEN MEMBERS OF THE SAME SPECIES BUT OF DIFFERENT GENETIC MAKEUP.

CERTAIN BODY LOCATIONS WHICH TOLERATE TRANSPLANTS AND WHERE NO IMMUNOLOGICAL ATTACK WILL OCCUR.

CHAPTER 26

GAS EXCHANGE

BEFORE READING ACTIVITIES

1. Since it is important to transfer information from your short-term to your long-term memory, you must review it. To review the information in chapter 25, use the following steps:
   1. Turn to the outline on page 522. Recite (to yourself or to someone else) the information associated with each point in the outline.
   2. If you can't remember what information goes with an idea, turn back to the appropriate text section and use your textbook marking.
   3. Review the figures in the text and test yourself to see if you can explain them without help from the verbal portions on the text.
   4. Test yourself on the vocabulary. Keep practicing terms you have not learned yet.
2. Turn to page 542 of the text and read the outline which lists the important concepts about "Gas Exchange." Answer the following questions.
   1. What are the three major concepts to be discussed

      1.   3.

      2.

   2. What are the adaptations for gas exchange?

      1.   3.

      2.   4.

   3. What are the main ideas you will study about the human respiratory system?

      1.   3.

      2.   4.

      5.

3. Read the Learning Objectives listed below and answer the questions which follow.

Learning Objectives	INFORMATION FOUND
After you have studied this chapter you should be able to:	
1. Compare various adaptations for gas exchange, including tracheal tubes, gills, lungs, and the body surface.	1. Figure 26-4
2. Compare air and water as sources of oxygen.	2. text 547-548

3. Trace the route traveled by a breath of air through the human respiratory system from nose to air sacs and, finally, to recipient cells. — 3. text 548-551, Figure 26-5

4. Describe breathing and its regulation. — 4. Figure 26-7, text 551-552

5. Compare the composition of exhaled air with that of inhaled air, and describe the exchange of oyxgen and carbon dioxide in the lungs and tissues. — 5. Table 26-1

6. Summarize the mechanisms by which oxygen and carbon dioxide are transported in the blood. — 6. ____________________

7. Describe the effects of breathing polluted air on the respiratory system. — 7. ____________________

1. In what part of the chapter will you probably get the information to help you compare the adaptations for gas exchange (Objective 1)? Put your answer in the "Information Found" Column.

2. What does "trace the route" mean? Where will you find the information you need to be able to show how air moves through the human body?

3. Read each of the objectives and go through the chapter to find where the necessary information is located.

4. Read the vocabulary terms and organize them into groups. Be sure these groups make sense to you. Some of the things you should notice as you read and organize the terms:
   1. Many common, everyday terms are used: respiration, book lungs, swim bladders, air sacs. Be sure you know what these terms mean when they relate to gas exchange.
   2. There are some sets of terms which are almost alike: pharynx and larynx, expiration and inspiration. How are these terms alike? How are they different?

5. Read the chapter introduction on page 543, the chapter summary on page 556 and use your knowledge of the vocabulary terms to take the Post-Test on pages 556 and 557.

DURING READING ACTIVITIES

1. Before you begin reading the section on "Adaptations for Gas Exchange" think of at least 2 questions you expect to find answers to while you read this section. Examples might be, "what is gas exchange?" "Who cares about gas exchange?" (or "Why is gas exchange important?"), or "What does body surface have to do with gas exchange?"

2. Use the BOLD FACE and Dark Face headings as guides. Each time you are ready to begin a new section, ask yourself some questions about the section and then read to find the answers to your own questions.

3. Continue to mark the text, paragraph by paragraph. When you have read and underlined a paragraph, write a summary of the paragraph in the margin. Be sure to summarize the paragraph in your OWN WORDS. To summarize, you must sort out the important from the unimportant or less important information.

4. Read the figures carefully. Some of the things to look for include:
   1. Note that Figure 26-3 has 5 separate parts which are all related to one topic, the function of the fish gill. What does each part depict? What does the whole figure depict?

   2. Figure 26-4 is a comparison chart which shows how lung structure varies in different vertebrate classes. Can you tell how each lung is different? How will this figure help you achieve the first learning objective?

   3. Figure 26-5 shows the upper body of a human being. Can you find and name each part? Can you explain what each part does? Do you know where these parts are in your own body? To practice labelling the various parts, use the diagram on page 557.

   4. What does Figure 26-7 show? It has two parts (a) & (b); how are these parts different? In what ways are parts (a) and (b) alike? In what way is the left portion of part (a) different from the right portion of part (a)?

   5. Note that Figure 26-8 is different from the previous figures. It is a flow chart which depicts how gas is exchanged. What is the difference between the solid and broken line arrows? Why do some arrows point in one direction and others in the opposite direction?

AFTER READING ACTIVITIES

1. Review and correct your textbook marking. Be sure that your summarizing notes in the margins are complete enough to make sense.

2. Review your vocabulary. Study terms in groups of 5-7 at any one time. Say or write the definition of each term before you turn the card over and look at what is written.

3. Answer the following questions:

   1. Distinguish between cellular respiration and organismic respiration.

2. What characteristics of the annelids, small arthropods and some mollusks enable these animals to get sufficient oxygen through their body surface?

3. What is the advantage of a countercurrent exchange system?

4. When the water levels get low, some fish are able to gulp air into their pharynx which than enters the ______________ __________ . Gas exchange occurs.

5. Gas exchange with _____________ is far more efficient than gas exchange with _______________. This is because __________ contains more oxygen than _____________.

6. Consider the human respiratory system, and list in sequence the structures through which air passes on its way to the lungs.

7. How much mucous is produced by the nasal epithelium in one day?

8. In humans, are the lung perfectly symmetrical structures? Explain.

9. Inside the lungs the _____________ branch extensively to form the _______________, these latter structures (more than a million in humans) each lead to tiny air sacs called _______________.

10. The mechanical process of moving air into and out of the lungs is called ___________________.

11. Study Figure 26-7. What causes air to rush into the lump?

12. The ____________________ ______________ is the vital link between the oxygen in the lump and the cells of the body.

13. Oxygen is more concentrated in the air sacs than in the lung _______________, so the oxygen moves into these structures. Carbon dioxide is more concentrated in the ________________ so it comes from these structures into the air sacs.

14. Examine Table 26-1 and answer the following questions.

    A. What gas is most prevalent in air?

    B. How much more carbon dioxide is there in exhaled air compared to inhaled air?

15. How is CO_2 transported in the blood?

Chapter 26 - Answers

1. Organismic respiration involves the uptake of oxygen by the organism, and its delivery to the cells. Cellular respiration refers to the metabolic pathways (glycolysis and the Krebs cycle) which produce energy.

2. These animals are small and have a low metabolic rate.

3. This system allows for more oxygen to diffuse into the blood. More efficient.

4. swim, bladder

5. air, water, air, water

6. Air enters through the nose, to the pharynx, through the larynx. into the trachea, the twin bronchi, and into the bronchioles of the lungs.

7. More than a pint. This means a healthy person swallows more than a pint of mucous a day.

8. No. The right lung has three lobes and the left lung has two lobes.

9. bronchi, bronchioles, alveoli

10. breathing

11. The diaphragm muscle contracts as do the rip muscles, and this causes an expansion of the chest cavity. This in turn causes the air to enter the lump.

12. circulatory system

13. capillaries, capillaries

14. A. Nitrogen

    B. Carbon dioxide is more than 100 times (140 times) more concentrated in exhaled air than in inhaled air.

15. About 20% is carried in the hemoglobin molecules, 10% is transported as CO_2, and the rest is transported as bicarbonate in the blood plasma.

ORGANISMIC RESPIRATION

RESPIRATION

VENTILATION

CELL RESPIRATION

GILLS

TRACHEAL TUBES

GILL FILAMENTS

OPERCULUM

LUNGS

COUNTERCURRENT EXCHANGE SYSTEM

AIR SACS

SWIM BLADDER

GAS EXCHANGE BETWEEN THE ORGANISM ON THE ENVIRONMENT.

OXYGEN FROM THE ENVIRONMENT IS TAKEN UP BY THE ORGANISM AND DELIVERED TO ITS CELLS, AND CARBON DIOXIDE IS EXCRETED TO THE ENVIRONMENT.

METABOLIC BREAKDOWN OF FUEL MOLECULES (E.G. GLUCOSE) WHICH REQUIRES OXYGEN AND YIELDS CARBON DIOXIDE AND ENERGY. GLYCOLYSIS AND THE KREBS CYCLE.

ANIMALS MOVE AIR OR WATER OVER THEIR CELLS OR RESPIRATORY STRUCTURES. MEANS OF BRINGING CONTINUOUS SUPPLY OF OXYGEN.

RESPIRATORY SYSTEM IN SOME ARTHROPODS AND SNAILS. AIR ENTERS THROUGH TINY OPENINGS CALLED SPIRACLES.

RESPIRATORY STRUCTURES FOUND MAINLY IN AQUATIC ANIMALS. OUTER SURFACE OF GILLS IS EXPOSED TO WATER AND THE INNER SIDE IS IN CONTACT WITH BLOOD VESSELS.

THE BONEY PLATE WHICH PROTECTS THE GILLS IN FISH.

STRUCTURES WHICH COMPRISE THE GILLS WHICH ARE SMALL AND WHICH CONTAIN BLOOD VESSELS SO THAT GAS EXCHANGE CAN OCCUR BETWEEN THE WATER AND THE ORGANISMS BLOOD.

ARRANGEMENT FOR GAS EXCHANGE IN GILLS IN WHICH THE BLOOD FLOWS IN THE OPPOSITE DIRECTION THAT THE BLOOD FLOWS.

RESPIRATORY STRUCTURES WHICH DEVELOP ARE INGROWTHS OF THE BODY SURFACE OR FROM THE WALL OF THE BODY CAVITY.

STRUCTURE HOMOLOGOUS TO LUNGS FOUND IN FISH. FUNCTION AS HYDROSTATIC ORGANS WHICH MAY STORE OXYGEN.

RESPIRATORY STRUCTURES IN BIRDS WHICH EXTEND INTO ALL PARTS OF THE BODY.

PHARYNX

NASAL CAVITIES

EPIGLOTTIS

LARYNX

TRACHEA

COUGH REFLEX

LUNGS

BRONCHI

PLEURAL CAVITY

PLEURAL MEMBRANE

ALVEOLI

BRONCHIOLES

CAVITY OF THE NOSE LINED WITH MOIST EPITHELIUM, AND THIS LINING HAS A RICH BLOOD SUPPLY TO WARM THE AIR.

THE BACK OF THE NASAL CAVITIES WHICH ARE CONTINUOUS WITH THE THROAT.

"ADAMS APPLE" OR "VOICE BOX."

FLAP OF TISSUE WHICH COVERS THE LARYNX SO NO FOOD OR LIQUID ENTERS THE AIR PASSAGES.

REACTION TO EXPEL FOOD OR LIQUID FROM THE RESPIRATORY SYSTEM.

TUBE STRUCTURE MAINTAINED BY RINGS OF CARTILAGE. LEADS FROM LARYNX TO THE BRONCHI. MUCOUS MEMBRANE LINED.

STRUCTURES WHICH LEAD FROM TRACHEA TO THE LUNGS.

LARGE PAIRED SPONGY ORGANS OCCUPYING THE CHEST CAVITY. RIGHT LUNG IS DIVIDED INTO THREE LOBES, THE LEFT LUNG IS DIVIDED INTO TWO LOBES.

MEMBRANE WHICH COVERS THE LUNGS. FORMS A CONTINUOUS SAC WHICH ENCLOSES THE LUNGS AND CONTINUES TO LINE THE CHEST CAVITY.

THE SPACE BETWEEN THE PLEURAL MEMBRANE AND THE MEMBRANE LINING THE CHEST CAVITY.

VERY SMALL BRANCHES OF THE BRONCHI.

TINY AIR SACS IN THE LUNG INTO WHICH EXTEND THE BRONCHIOLES. GAS EXCHANGE OCCURS IN THESE STRUCTURES.

DIAPHRAGM

BREATHING

ASTHMA

OXYHEMOGLOBIN

THE MECHANICAL PROCESS OF MOVING AIR FROM THE ENVIRONMENT INTO THE LUNGS, AND EXPELLING THE AIR FROM THE LUMGS. AVERAGE ADULT BREATHES TWELVE TIMES PER MINUTE.

THE MUSCULAR FLOOR OF THE CHEST CAVITY.

HEMOGLOBIN COMBINED WITH OXYGEN.

CHRONIC BRONCHIAL CONSTRICTION SOMETIMES CAUSED BY ALLERGIES.

CHAPTER 27

PROCESSING FOOD

BEFORE READING ACTIVITIES

1. Before you begin to read the chapter about processing food, review the information about gas exchange. Remember that gas exchange is one important process that animals must have to survive and that food processing is another process important for animal survival. To review, use the following steps:

    1. Turn to page 542 and try to do what is asked in each of the objectives. Either write your answers or say your answers aloud.

    2. Supplement what you remember by using your textbook marking.

    3. Review the figures in the textbook. Explain each figure without using the verbal portion of the text.

    4. Test yourself on the vocabulary.

2. Turn to page 558 of the textbook and read through the outline about processing food. Answer the following questions.

    1. What are "dinner jackets?" Why are there different kinds of dinner jackets?

    2. Why are there separate sections about the digestive systems of invertebrate animals and vertebrate animals?

    3. What are the important ideas discussed about the vertebrate digestive system?

        1.

        2.

        3.

        4.

        5.

        6.

        7.

        8.

        9.

3. Read the Learning Objectives listed below and answer the questions which follow.

KEY WORD	LEARNING OBJECTIVES	INFORMATION FOUND
	After you have read this chapter, you should be able to:	
1. describe	1. Describe in general terms the following steps in processing food: ingestion, digestion, absorption, and elimination.	1. Introduction, p. 559
2. compare	2. Compare adaptations that herbivores, and omnivores possess for their particular mode of nutrition.	2. pages 559-562
3. compare	3. Compare the nutritional life-styles of parasites, commensals, and mutualistic partners.	3. pages 561-562
4. contrast	4. Contrast the digestive systems of cnidarians, flat-worms, and earthworms.	4. Figure 27-4
5. Identify and give	5. Identify on a diagram or model each of the structures of the human digestive system described in this chapter, and give the function of each structure.	5.
6.	6. Trace the pathway traveled by an ingested meal, describing each of the changes that takes place en route. (Your instructor may ask you to describe in sequence the step-by-step digestion of carbohydrates, lipids, and proteins.)	6.
7.	7. Describe the protective mechanisms that prevent gastric juice from digesting the stomach wall, and explain what happens when these mechanisms fail.	7.
8.	8. List the functions of the liver and pancreas.	8.
9.	9. Draw and label a diagram of an intestinal villus, and describe its function.	9. Figure 27-9

1. Underline the KEY WORD which tells you what you will need to be able to do to fulfill each objective. Be sure to read the whole objective because there may be more than one key word. (See Objective 5.)

2. Write the key word or key words for each objective in the column labelled "key word."

3. Determine which part of the chapter will discuss the information needed to fulfill the objective. Write the page, table, or figure in the column labelled "information found."

4. Read through the vocabulary terms and organize them into groups. Some of the things you should think about include the following:

   1. Which terms are about the same idea? For example, the following terms are all symbionts: parasite, ectoparasite, endoparasite, commensal, and mutualistic partners. You should probably group all of the terms about the liver together and all of the terms about the pancreas together. What other groups can you find in this new vocabulary?

   2. Check to see which terms share word parts. Examples include digestion and egestion and ectoparasite and endoparasite.

5. Read the chapter introduction on page 559, the summary on pages 572 and 573, and then take the Post-Test on pages 573 and 574.

DURING READING ACTIVITIES

1. Remember to ask yourself questions using the topic headings before you begin to read each new section. Your first questions should be about "dinner jackets." Your next questions should be about the adaptations of herbivores. Then, read to find the answers to your questions.

2. As you read, look for the information to fulfill the objectives listed at the beginning of the chapter. Think about what each of the Key Words asked you to do.

   1. Objectives 2 and 3 asked you to COMPARE things. Compare means to show both the SIMILARITIES AND the DIFFERENCES. This means that the information you need must explain that herbivores, carnivores and omnivores have some adaptations that are alike and some that are not alike. You must first tell what the adaptations are for each group and then tell whether these adaptations are similar to or different from the adaptations of the other groups.

   2. Objective 4 asks you to CONTRAST the life-styles of three things. This means that you must give information that explains what each life style is and how these life styles are DIFFERENT.

   3. Objective 5 asks you to be able to identify the structures of the human digestive system. To do this, you need to find the appropriate information in the text. This information is available both in verbal and pictorial form. Use the figure (27-5) to practice finding the anatomical parts of the body. Be sure you can explain the function of each part. To practice on a diagram which does not have labels, use the one given in the Post-Test.

3. You should be marking the text as you read. Check the section of the text below (from pages 561 and 562) to see how your marking compares.

SYMBIONTS

A symbiont is an organism that lives in intimate association with a member of another species. One or both of the organisms usually derive nutritional benefit from the association. Three types of symbionts are (1) parasites, (2) commensals, and (3) mutualistic partners.

A parasite lives in or on the body of a living plant or animal, the host species, and obtains its nourishment from the host (Fig. 27-3). Since the parasite's environment is its host, an effective parasite does not actually kill its host. For example, a tapeworm might live within the intestine of its host for many years. Ectoparasites such as fleas and ticks, live outside the host's body; endoparasites, such as tapeworms and hookworms, live inside the host. Whether the parasite nourishes itself from food ingested by its host or by sucking the host's blood, it is strictly a freeloader.

1. Did you circle the new vocabulary and underline the definition?

2. Did you number the types of symbionts in the body of the text?

   Be sure you do this so when you review you will realize how many parts there are in any one topic.

3. Did you summarize the information in the margin? Your summary should use your own symbol system; you need to use the same symbol system in all of your marking so that you will understand what you mean when you use your notes for review.

4. There are two tables (27-1 and 27-2) which contain important information. How can you reorganize this information so it will be easier to learn? For example, you might learn about carbohydrate digestion at one time, protein digestion at another time, and lipid digestion some time later. In addition, you might want to make study cards to help you practice this information.

AFTER READING ACTIVITIES

1. Review and correct your textbook marking.

2. Go back to the objectives on page 558 and write down or say aloud the information required to fulfill each objective.

3. Answer the following questions.

   1. List the four steps of food processing.

   2. Herbivores are animals that consume __________ as their source of food. This food contains a large amount of ____________ which is not digested by the animal.

   3. Animals that consume meat are _____________, and therefore are frequently _______________.

   4. What are the three types of symbionts?

   5. Give two examples of ectoparasites and two examples of endoparasites.

6. Give an example of a mutualistic partnership.

7. Give an example of a commensal organism.

8. _______________ ____________ are two species of organisms which live together for their mutual benefit.

9. List in order the organs through which food passes during digestion.

10. Name three accessory glands found in humans.

11. How much saliva is produced in one day by an average adult?

12. Food is moved along the digestive tract by ____________ __________. Another factor which helps move the food in the digestive tract is _______________.

13. Examine Figure 27-8

    A. How many different cell types are identified in the gastric glands? What are they?

    B. What is produced by each of these three different cells?

14. Where does most of the chemical digestion of food occur?

15. ______________ from the liver and _______________ from the pancreas are released into the duodenum to act upon the partially digested food.

16. The final steps of digestion involve the utilization of enzymes produced by the _______________ cells of the ______________________________.

17. What is the function of the intestinal villi?

18. List some of the major functions of the liver.

19. Examine Table 27-1 and answer the following questions.

A. In which organs does lipid digestion occur?

B. What organs produce the enzymes for protein digestion in the stomach and small intestine?

C. In which organs does carbohydrate digestion occur?

20. Examine Figure 27-9 and answer the following questions.

A. On what structures would you expect to find the microvilli?

B. What structures are present in the villi to carry the digested nutrients?

21. Examine Figure 27-11. Which structure carries bile from the gall bladder to the duodenum?

22. Examine Table 27-2. Which two systems are involved in the regulation of digestion?

23. Name two factors which can increase the probability of one getting an ulcer.

Chapter 27 - Answers

1. Ingestion, digestion, absorption, and elimination.

2. plants, cellulose

3. carnivores, predators

4. Parasites, commenseals, and mutualistic partners.

5. Ectoparasites include fleas and ticks and endoparasites include hook worms and tape worms.

6. Termites which eat the wood and their gut flagellates which produce the enzymes to digest the wood.

7. Every worm burraut or hermit crab shell has other organisms living in it.

8. Mutualistic partners

9. mouth, pharnyx, esophagus, stomach, small intestine, large intestine, and anus.

10. Pancreas, liver, salivary glands

11. approximately one liter.

12. peristaltic motion, gravity

13. A. Three different cell types. Parietal cells, mucous cells, and chief cells.

    B. Parietal cells - hydrochloric acid
    Mucous cells - Mucous
    Chief cells - Pepsin or pepsinogen

14. Duodenum (<u>not</u> the stomach)

15. Bile, enzymes

16. epithelial, lining of the duodenum.

17. They increase the surface area of the intestine for greater absorption of nutrients.

18. A. Secretes bile for digestion of fats

    B. Monitors absorbed nutrients in blood

    C. Converts excess glucose to glycogen

    D. Stores iron and certain vitamins

    E. Manufactures some blood proteins.

    F. Converts excess amino acids to fatty acids and urea.

    G. Detoxifies many drugs and poisons

    H. Removes bacteria and worn-out red blood cells from the circulation.

19. A. Small intestine

    B. Stomach gastric glands, pancreas, and small intestine.

    C. Mouth, stomach, small intestine.

20. A. The villi

    B. The capillary network and the lymphatic system.

21. The common bile duct

22. Hormonal (endocrine) system and the nervous system.

23. stress, smoking, alcohol, aspirin.

NUTRITION

NUTRIENTS

DIGESTIVE SYSTEM

HETEROTROPHS

DIGESTION

INGESTION

ELIMINATION

ABSORPTION

CARNIVORES

HERBIVORES (PRIMARY CONSUMERS)

SYMBIONT

OMNIVORES

SUBSTANCES PRESENT IN FOOD WHICH ARE NEEDED BY AN ORGANISM AS AN ENERGY SOURCE, TO MAKE COMPOUNDS FOR METABOLIC PROCESSES, AND AS BUILDING BLOCKS TO MAKE MOLECULES FOR GROWTH AND REPAIR.

THE PROCESS BY WHICH AN ORGANISM TAKES IN AND ASSIMILATES FOOD.

ALL ANIMALS. ORGANISMS THAT MUST OBTAIN THEIR ENERGY AND NOURISHMENT FROM MOLECULES MANUFACTURED BY OTHER ORGANISMS.

ORGANISMS SYSTEM THAT PROCESSES THE FOOD THEY EAT. CONSISTS OF INGESTION, DIGESTION, ABSORPTION, AND ELIMINATION.

THE TAKING OF FOOD INTO THE BODY AFTER SELECTION OF THE FOOD.

THE PROCESS BY WHICH THE INGESTED FOOD IS MECHANICALLY BROKEN DOWN INTO SMALLER PIECES, AND THEN ENZYMATICALLY DEGRADED INTO ITS COMPONENTS.

IN ORGANISMS WITH INTRACELLULAR DIGESTION IT IS THE PASSAGE OF NUTRIENTS FROM THE FOOD VACUOLE INTO THE CYTOPLASM. IN ANIMALS WITH DIGESTIVE TRACTS IT REFERS TO PASSAGE OF NUTRIENTS FROM CELLS OF THE DIGESTIVE TRACT TO THE BLOOD.

FOOD THAT IS NOT DIGESTED AND ABSORBED IS DISCHARGED FROM THE BODY.

ANIMALS WHICH EAT ONLY PLANT MATERIAL.

ANIMALS THAT EAT FLESH.

ANIMALS THAT CONSUME BOTH PLANT MATERIAL AND MEAT.

AN ORGANISM THAT LIVES IN CLOSE ASSOCIATION WITH A MEMBER OF ANOTHER SPECIES. ONE OR BOTH ORGANISMS DERIVE NUTRITIONAL BENEFIT FROM THE ASSOCIATION.

ECTOPARASITES

PARASITE

COMMENSAL

ENDOPARASITES

COMPLETE DIGESTIVE SYSTEM

MUTUALISTIC PARTNERS

VERTEBRATE DIGESTIVE SYSTEM

PERISTALTIC CONTRACTIONS

BOLUS

SALIVARY AMYLASE

EPIGLOTTIS

PHARYNX

ORGANISM WHICH LIVES IN OR ON THE BODY OF A LIVING PLANT OR ANIMAL, THE HOST SPECIES, AND OBTAINS NOURISHMENT FROM THE HOST. DOES NOT KILL THE HOST.

PARASITES WHICH LIVE OUTSIDE THE HOST'S BODY.

PARASITES WHICH LIVE INSIDE THE HOST'S BODY.

ORGANISM THAT DERIVES BENEFIT FROM ITS HOST WITHOUT HARM OR BENEFIT TO THE HOST.

TWO SPECIES OF ORGANISMS THAT LIVE TOGETHER FOR THEIR MUTUAL BENEFIT. MAY BE ABLE TO SURVIVE SEPARATELY.

THE DIGESTIVE TRACT IS A COMPLETE TUBE WITH A MOUTH THROUGH WHICH FOOD ENTERS AND AN ANUS THROUGH WHICH UNDIGESTED FOOD IS ELIMINATED.

WAVES OF MUSCULAR CONTRACTION WHICH PUSH THE FOOD IN ONE DIRECTION. ALLOWS FOR MORE FOOD TO BE TAKEN IN WHILE OTHER FOOD IS BEING DIGESTED AND ABSORBED.

COMPLETE SYSTEM. CONSISTS OF: MOUTH, PHARYNX (THROAT), ESOPHAGUS, STOMACH, SMALL INTESTINE, LARGE INTESTINE, AND ANUS. VERTEBRATES HAVE ACCESSORY GLANDS WHICH SECRETE DIGESTIVE JUICES INTO THE DIGESTIVE TRACT.

AN ENZYME PRODUCED BY THE THREE PAIRS OF SALIVARY GLANDS WHICH BEGINS THE DIGESTION OF CARBOHYDRATES.

A LUMP OF CHEWED, SWALLOWED FOOD.

MUSCULAR TUBE THAT LEADS TO BOTH THE RESPIRATORY AND THE DIGESTIVE SYSTEMS.

THE SMALL FLAP OF TISSUE WHICH COVERS THE OPENING TO THE AIRWAY WHEN FOOD IS SWALLOWED.

RUGAE

STOMACH

CHIEF CELLS

GASTRIC GLANDS

PARIETAL CELLS

PEPSIN

DUODENUM

CHYME

MICROVILLE

VILLI (VILLUS, SINGULAR)

LIVER

BRUSH BORDER

LARGE, MUSCULAR ORGAN WHICH HAS AN INNER LINING THAT IS HIGHLY FOLDED. HAS ONE QUART CAPACITY.

FOLDS OF THE INNER STOMACH WALL.

GLANDS IN STOMACH WALL.

CELLS IN THE GASTRIC GLANDS WHICH SECRETE PEPSINOGEN WHICH IS THEN CONVERTED TO PEPSIN.

THE MAIN DIGESTION ENZYME IN THE STOMACH.

CELLS OF THE GASTRIC GLANDS WHICH SECRETE HYDROCHLORIC ACID.

THE MASHED, CHURNED, PARTIALLY DIGESTED FOOD FOUND IN THE STOMACH.

THE FIRST PORTION OF THE SMALL INTESTINE WHERE MOST DIGESTION OCCURS.

TINY FINGER-LIKE PROJECTIONS FROM THE WALL OF THE INTESTINE WHICH INCREASE THE SURFACE AREA FOR ABSORPTION OF NUTRIENTS.

VERY SMALL CYTOPLASMIC PROJECTIONS FROM THE SURFACE OF THE INTESTINAL EPITHELIAL CELLS. ABOUT 600 MICROVILLI PROJECT FROM EACH EPITHELIAL CELL.

THE "FUZZY" LAYER FORMED BY THE MICROVILLI ON THE EPITHELIAL CELLS.

THE LARGEST AND MOST COMPLEX ORGANS IN THE BODY. A SINGLE LIVER CELL CAN CARRY OUT MORE THAN 500 METABOLIC ACTIVITIES.

EMULSIFICATION

GALLBLADDER

TRYPSIN AND CHYMOTRYPSIN

PANCREAS

PANCREATIC AMYLASE

PANCREATIC LIPASE

CECUM

RIBONUCLEASE AND DEOCYRIBONUCLEASE

PEPTIC ULCER

VERMIFORM APPENDIX

STRUCTURE IN WHICH BILE PRODUCED IN THE LIVER IS STORED.

MECHANICAL DIGESTION OF FATS WHERE THE BILE CAUSES THE FATS TO BE BROKEN INTO SMALLER PARTICLES.

LARGE GLAND THAT SECRETES DIGESTIVE ENZYMES AND HORMONES WHICH REGULATE BLOOD GLUCOSE LEVELS

PANCREATIC ENZYMES WHICH DIGEST POLYPEPTIDES TO DIPEPTIDES

PANCREATIC ENZYME WHICH DIGESTS FATS

PANCREATIC ENZYME WHICH BREAKS DOWN MOST CARBOHYDRATES

PANCREATIC ENZYMES WHICH BREAKDOWN NUCLEIC ACIDS TO NUCLEOTIDES

BLIND POUCH CREATED BY JUNCTION OF THE SMALL AND LARGE INTESTINE

PROJECTION FROM END OF CECUM

OPEN SORE IN THE STOMACH CAUSED BY GASTRIC ACIDS DIGESTING A PORTION OF THE STOMACH

CHAPTER 28

NUTRITION

BEFORE READING ACTIVITIES

1. Before you begin to read the chapter about Nutrition, review the information in the chapter about Processing Food. Each of these chapters discusses concepts that are closely related. To review, use the following steps:

   1. Turn to page 558 and try to explain what each of the headings in the outline means. Either write your answers or say your answers aloud.

   2. Review your vocabulary terms. The terms used in this chapter will probably be used in the chapter about nutrition.

   3. Review the information in the Tables and Figures. Be sure you can explain each figure without rereading the verbal description.

2. Turn to page 575. Read the outline about nutrition. Answer the following questions.

   1. What are the basic nutrients?

      1.   4.

      2.   5.

      3.   6.

   2. What are the major areas you need to understand about metabolism?

      1.   3.

      2.   4.

   3. What should you think about when you discuss the problem of world food supply?

      1.   3.

      2.

3. To make learning easier, it is helpful to see the relationships among ideas, to group ideas into categories, and to relate new ideas to what we already know. One way to see relationships among ideas is to visualize how the ideas fit together. Constructing a graphic organizer is one form of picturing the relationship among ideas. An easy way to begin a graphic organizer is to use the outline given at the beginning of the chapter. The following "picture" is a graphic organizer for chapter 28.

NUTRITION

* Carbohydrates

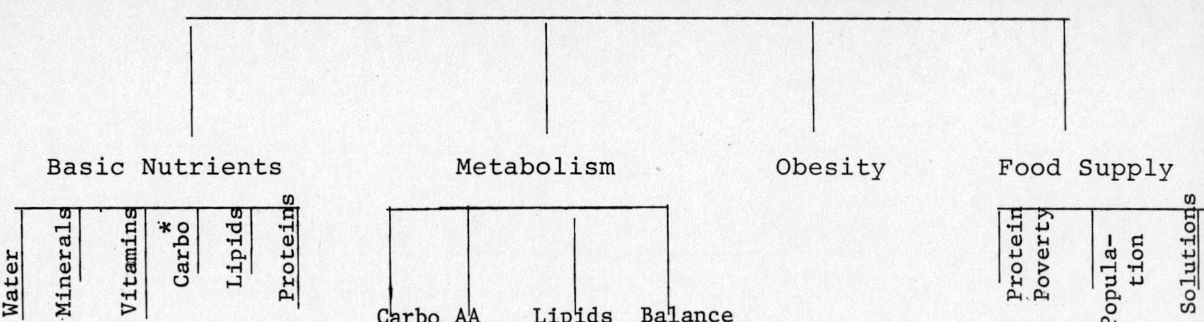

Can you see how the outline was used to construct this graphic organizer? What sub-topics should go under the solution section?

4. Read through the vocabulary terms. Note that there are not very many new terms in this chapter. This is a good time to review terms from the chapters you have already studied.

5. Read the introductory paragraph on page 576 (short as it is), the summary on pages 591 and 592, and take the Post-Test on page 592.

<u>DURING READING ACTIVITIES</u>

1. Five of the Learning Objectives ask you to DESCRIBE something (the effect of nutrient deficiency, the fate of glucose, etcetera). Describe means to RETELL information giving the IMPORTANT POINTS. Describing is similar to summarizing. Steps to use while you read should include:

   1. Mark the main ideas of each paragraph.

   2. Mark the <u>most important details</u>. Use some system to differentiate between the main ideas and the important details. You might asterisk main ideas or underline main ideas twice or use a curved line under main ideas. You can't remember everything, so you <u>must differentiate</u> between more important and less important ideas.

   3. When you have finished reading from one subheading to the next one, write your summary or description of the concept presented.

2. As you read, think about important points that should be included in the lower levels of your graphic organizer.

3. In this chapter, you are asked to DISCUSS the solutions to...(see objective 10). DISCUSS means that you must EXPLAIN what an idea, concept, or theory is by CHOOSING DETAILS about that idea or concept.

<u>AFTER READING ACTIVITIES</u>

1. Review your textbook marking. If you have not marked enough informatio so that it still makes sense when you review, be sure to correct your marking. If you have marked <u>too much</u>, correct this by using a second type of marking. If you <u>highlighted</u>, go back and underline the important parts only; if you underlined, go back and highlight the important parts.

2. Test yourself on the new vocabulary.

3. Review the information in Table 28-1 and Table 28-2. Think of how each bit of information relates to your own body. What will happen if you do not get enough calcium? or enough Vitamin C?

4. Go back to the graphic organizer at the beginning of this chapter. Try to explain to yourself what concepts are involved in each of the headings and subheadings. Add as many lower level details as you can.

5. Answer the following questions.

   1. What are the essential nutrients that may be supplied by a well-balanced diet?

   2. What is the daily intake of water for an adult human?

   3. The human body is approximately 65% ____________.

   4. What minerals are needed in the largest quantity?

   5. Phosphorus and _____________ are important in the structure of bones and teeth.

   6. ______________ is the mineral most likely to be lacking in the diet.

   7. Examine Table 28-2 and answer the following questions.

      A. Name the fat soluble vitamins.

      B. How many different vitamins are there in the B-complex group of vitamins?

   8. Moderate overdoses of vitamin B and C are ____________, but moderate overdoses of the fat-soluble vitamins are not easily ____________.

   9. Which of the food types makes up the largest proportion of the American diet?

   10. Name some foods rich in carbohydrate.

   11. In general, plant foods are rich in ________________ fats and no ________________, while animal foods are rich in _____________ fats and ___________________.

   12. Diets high in _______________ are likely to be more expensive than diets rich in _____________________.

13. Ingested proteins are digested to their molecular units, the ____________ ____________. These molecules are then reutilized by the cells to make the ____________ of the cell.

14. In the wealthier, more developed countries in the world, the dietary intake of ________________ is far greater than necessary.

15. Most humans in the world depend on the ____________ __________ as the staple food. Protein content of these foods is inadequate.

16. The regulation of blood glucose levels is carried out by the ____________ cell. All cells, and especially brain cells are dependent on a continual supply of ____________.

17. When a person eats too much carbohydrate, and the liver glycogen levels are high, than the excess carbohydrate is converted to __________ ___________.

18. Examine Figure 28-2, and answer the following questions.

    A. When there is no glucose or glycogen available, the cells enter into the process of ________________________.

    B. When glucose levels are high, the excess glucose in the liver is converted to ___________________ by the process of __________________.

19. People tend to gain weight when they take in ___________ energy in food than they expend in daily activity.

20. According to insurance company statistics, men who are 20% or more overweight have an increased rate of ___________ ___________, ___________ hemorrhage, and ______________.

21. Every year _________ ___________ people die as a result of malnutrition.

22. Children who are malnourished during their early years may have ____________ ______________.

23. In order to adequately feed the world's population, the ___________ ____________ must be increased.

24. The main problem with respect to the food supply, is that there are two many ________________.

Chapter 28 - Answers

1. Proteins, carbohydrates, lipids, vitamins, minerals, and water.
2. 2.5 quarts (2.4 liters)
3. water.
4. sodium chloride
5. calcium
6. iron
7. A. A, D, E, K

   B. There are eight.
8. excreted, excreted
9. carbohydrates
10. Rice, potatoes, corn, and other cereal grains
11. polyunsaturated, cholesterol, saturated, cholesterol
12. protein, carbohydrates
13. amino acids, proteins
14. protein
15. cereal grains
16. liver, glucose
17. body fat
18. A. gluconeogenesis

    B. glycogen, glycogenesis
19. more
20. heart disease, cerebral, diabetes
21. 15 million
22. brain damage
23. food supplies
24. people

TRACE ELEMENTS

MINERALS

VITAMINS

IRON-DEFICIENCY ANEMIA

WATER-SOLUBLE VITAMINS

FAT-SOLUBLE VITAMINS

LIPIDS

CARBOHYDRATES

ESSENTIAL AMINO ACIDS

PROTEINS

GLYCOGENOLYSIS

GLYCOGENESIS

INORGANIC NUTRIENTS INGESTED AS SALTS DISSOLVED IN FOOD AND WATER.

INORGANIC NUTRIENTS WHICH ARE REQUIRED IN MINUTE QUANTITIES.

CONDITION WHERE PEOPLE LACK ENERGY. DUE TO INADEQATE DIETARY INTAKE OF IRON WHICH IS ESSENTIAL TO MAKE HEMOGLOBIN FOR OXYGEN TRANSPORT.

ORGANIC COMPOUNDS REQUIRED BY THE BODY FOR BIOCHEMICAL PROCESSES. SAME FUNCTION AS COENZYMES.

THOSE VITAMINS WHICH CAN BE DISSOLVED IN FATS AND OILS. INCLUDES VITAMINS A, D, E, AND K.

THOSE VITAMINS WHICH DISSOLVE IN WATER. INCLUDES B VITAMINS AND VITAMIN C.

BODY'S PRINCIPAL FUEL. SUGAR, STARCH AND CELLULOSE.

WATER INSOLUBLE CARBON-RICH MOLECULES. LIPID FUNCTION AS FUEL MOLECULES, COMPONENTS OF THE CELL MEMBRANE, AND CHOLESTEROL.

CRITICAL AS NUTRIENTS BECAUSE THEY ARE ESSENTIAL BUILDING BLOCKS OF CELLS. THE MOST ABUNDANT MATERIAL IN BODY SOLIDS.

OF THE 20 DIFFERENT AMINO ACIDS, 8 (9 IN CHILDREN) CAN NOT BE MADE IN THE BODY AND MUST BE PROVIDED BY THE DIET.

THE FORMATION OF GLYCOGEN IN THE LIVER. GLYCOGEN IS A LARGE MOLECULE COMPOSED OF MANY GLUCOSE MOLECULES LINKED TOGETHER.

THE ORDERLY BREAKDOWN OF GLYCOGEN TO YIELD GLUCOSE IN RESPONSE TO FALLING BLOOD GLUCOSE LEVELS.

DEAMINATED

GLUCONEOGENESIS

β (BETA) OXIDATION

UREA

BASAL METABOLIC RATE (BMR)

METABOLISM

TOTAL METABOLIC RATE

THE FORMATION OF GLYGOGEN FROM AMINO ACIDS AND GLYCEROL AFTER THE GLYCOGEN STORES HAVE BEEN USED UP.

WHEN AMINO ACIDS ARE METABOLIZED IN THE LIVER, THE AMINE GROUP IS REMOVED FROM THE AMINO ACID.

NITROGEN-CONTAINING EXCRETION FROM THE BODY.

THE LIVER PROCESS IN WHICH THE FATTY ACIDS ARE BROKEN DOWN TO SMALL UNITS AND COMBINED WITH COENZYME A TO FORM ACETYL COENZYME A, A FUEL MOLECULE.

ALL THE CHEMICAL REACTIONS WHICH OCCUR IN THE BODY.

THE BASIC COST OF METABOLIC LIVING. THE RATE OF ENERGY UTILIZATION DURING RESTING CONDITIONS.

THE SUM OF A PERSONS BMR AND THE ENERGY USED TO CARRY OUT DAILY ACTIVITIES.

CHAPTER 29

DISPOSAL OF METABOLIC WASTES

BEFORE READING ACTIVITIES

1. Before you begin to read, review the important concepts from chapter 28. What are the basic nutrients? What does each do? What concepts are associated with metabolism? What is obesity and how does it effect health? What are the problems of world food supply? What are some of the possible solutions?

2. Turn to page 594. Read the chapter outline about disposal of metabolic wastes. Answer the following questions.

   1. How many major concepts are presented in this chapter? (Hint: How many Roman numerals are there?)

   2. Which major concepts have minor ideas that you need to know about?

   3. Which is a broader concept, "The mammalian urinary system" or "Design of the urinary system?" How can you determine this?

   4. Construct a graphic organizer to show the relationship between major and minor ideas. Use another sheet of paper and turn it sideways so you can leave more room between ideas.

DISPOSAL OF METABOLIC WASTES

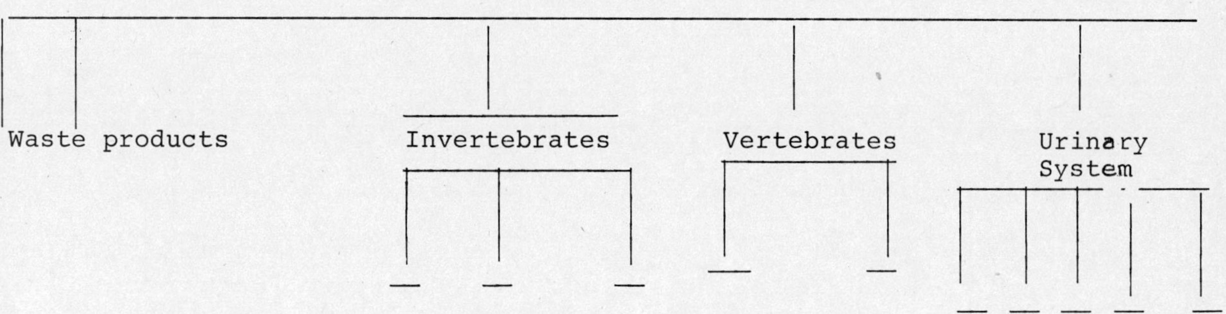

1. The basic framework for your organizer should look like the one above.

2. Fill in the missing key ideas for each line.

3. Use a different color of ink for each level of key ideas. This color coding will help you visualize which groups of ideas belong together.

3. Read the short introductory paragraphs on page 595. Read the summary on pages 607 and 608. Separate and read through the vocabulary cards. Use the information from these sources to take the Post-Test on page 608.

4. Read the Learning Objectives on page 594.

    1. As you read these objectives, underline the key words that tell what you need to able to do: describe, summarize, define, draw and label.

    2. Think about what each of these terms means.

    3. Go through the text to find out which pages will give you the information you need to accomplish each objective. Write the page numbers next to the objective.

DURING READING ACTIVITIES

1. Remember, you need to be able to summarize or describe ideas. Both of these key words ask you to give the important points only. This means you must think about what you read, find the main ideas, and eliminate the details.

2. You are asked to DEFINE some terms. When you define terms, you should give the MAJOR CHARACTERISTICS of the term. You should make your definition as concise as possible, but be sure to include all of the important ideas about the term.

3. In this set of Learning Objectives, you are asked to draw and label (see objectives 5 and 6). To do this you should use the following steps:

    1. First read the Figures (29-5 and 29-8) that give the information you need.

    2. Cover up the labels and try to put a label on each part. One way you can practice is to copy the outline (by hand or machine) and eliminate the verbal portion so you can test yourself. Number the lines and write your answers on a separate piece of paper so you can use the outline again.

4. Compare Figure 29-10 with Figure 29-5 and Figure 29-8. How are 29-5 and 29-8 alike? How do they differ from 29-10? What do the solid arrows mean? What do the broken line arrows mean?

5. Continue to mark your textbook.

    1. Differentiate between main ideas and minor ideas.

    2. Circle new vocabulary and underline the definitions.

    3. Use asterisks to indicate ideas of special significance.

    4. If you do not understand something, put a ? in the margin and then ASK THE INSTRUCTOR what the portion of the text means.

AFTER READING ACTIVITIES

1. Review your textbook marking. Make any needed corrections so you will NOT REREAD the text when you study for an examination.

2. Test yourself on the new vocabulary. Since there were not many new terms, take the opportunity to review vocabulary from several earlier chapters. If you don't continue to review new terms and use them, you will forget them.

3. Go back to the graphic organizer you made before you began reading this chapter. Explain to yourself or to another person what concepts are involved in each of the headings and subheadings. Fill in as many details as you can.

4. Answer the following questions.

    1. It is critical for organisms to be able to regulate the levels of fluid content, this is ____________________________.

    2. What are the metabolic wastes of most animals?

    3. How do simple organisms get rid of their metabolic wastes?

    4. The excretory system of insects are the ______________ ______________.

    5. The excretory system of vertebrates is more complicated than that seen in invertebrates, and consists of the ______________ ______________, the skin, __________ or gills, and the digestive system.

    6. How is it possible for some marine organisms to drink sea water?

    7. Marine mammals consume large amounts of salt water along with their food. How do they get rid of these salts?

    8. Examine Figure 29-4 and answer the following questions.

        A. Name the organs or organ systems involved in the excretion of water?

        B. Aside from water, what other wastes are removed by the lungs?

    9. List the components of the mammalian urinary system.

    10. Why do human males get fewer bladder infections than females?

11. List the major components of the nephron?

12. What is the fate of the useful substances that are initially filtered out of the blood?

13. What are the two phases of urine formation?

14. A rather large portion of the total plasma is forced out of the capillaries into Bowman's capsule. What percentage of the total plasma volume is processed in this way?

15. How long does it take for a volume of blood equal to the total body blood volume to pass through the kidneys?

16. How much glomerular filtrate is produced per day?

17. Is kidney filtration, a selective process? Explain.

18. What happens if the renal threshold for a particular substance is exceeded?

19. What is the composition of urine?

20. Is urine a clean fluid?

Chapter 29 - Answers

1. osmoregulation

2. Water, carbon dioxide, and nitrogenous wastes.

3. Diffusion

4. Malphigian tubules.

5. Urinary system, lungs

6. They retain the water, and the salts are selectively removed.

7. Their kidneys produce very concentrated urine that has a salt concentration that is higher than sea water.

8. A. Lungs, skin, digestive system, and kidney

   B. Carbon dioxide

9. The kidneys, ureters, and the urinary bladder.

10. Because the lengthy urethra in males prevents bacterial invasion, and females have a shorter urethra.

11. Bowman's capsule, renal tubule, and glomerulus.

12. These substances are returned to the blood.

13. Filtration through Bowman's capsule and resorption by the tubules.

14. Ten percent

15. Four minutes

16. 45 gallons per day

17. No, filtration is not a selective process, but vital materials such as glucose, amino acids, salts, and water are reclaimed by reabsorption.

18. The portion of a substance that is not resorbed is excreted with the urine.

19. 96% water, 2.5% nitrogenous wastes, and 1.5% salts and traces of other substances.

20. Urine as it comes out of the body is very clean, for it has been sent through an elaborate filter, the kidney.

EXCRETORY SYSTEM

EXCRETION

AMMONIA

OSMOREGULATION

NEPHRIDIAL ORGANS

UREA OR URIC ACID

METANEPHRIDIA

PROTONEPHRIDIA (FLAME CELLS)

MALPIGHIAN TUBULES

ANTENNAL GLANDS (GREEN GLANDS)

URINARY SYSTEM

KIDNEY

THE REMOVAL OF METABOLIC WASTES FROM THE BODY. IF ALLOWED TO ACCUMULATE, THEY WOULD BECOME TOXIC.

ELABORATE SYSTEM FOR THE REMOVAL OF WASTE FROM THE BODY. MAINTAIN HOMEOSTASIS BY SELECTIVELY ADJUSTING THE CONCENTRATION OF SALTS AND OTHER SUBSTANCES IN BLOOD AND BODY FLUIDS.

THE ABILITY OF AN ORGANISM TO REGULATE ITS FLUID CONTENT.

NITROGENOUS SUBSTANCE FORMED DURING THE METABOLIC BREAKDOWN OF AMINO ACIDS AND NUCLEIC ACIDS. HIGHLY TOXIC AND MUST BE EXCRETED.

NITROGEN-CONTAINING WASTE TO WHICH AMMONIA IS CONVERTED. IT IS LESS TOXIC THAN AMMONIA.

COMMON TYPE OF EXCRETORY ORGAN IN INVERTEBRATES. SIMPLE OR BRANCHING TUBES

PART OF EXCRETORY SYSTEM OF FLATWORMS

NEPHREDIAL ORGANS OF ANNELIDS-ROUNDWORMS

THE PRINCIPAL EXCRETORY ORGANS OF CRUSTACEANS.

THE EXCRETORY SYSTEM OF INSECTS.

THE MAIN EXCRETORY ORGAN IN VERTEBRATES. EXCRETES NITROGENOUS WASTES, WATER AND SALTS MAINTAINS INTERNAL CHEMICAL BALANCE OF THE BODY.

EXCRETORY SYSTEM WHICH CONSISTS OF THE KIDNEYS, THE URINARY BLADDER AND ASSOCIATED DUCTS.

MEDULLA OF KIDNEY

CORTEX OF KIDNEY

URETERS

RENAL PELVIS

URETHRA

URINARY BLADDER

NEPHRONS

URINATION

RENAL TUBULE

BOWMAN'S CAPSULE

PODOCYTES

GLOMERULUS

OUTER PORTION OF THE KIDNEY.	THE INNER PORTION OF THE KIDNEY
THE PORTION OF THE KIDNEY WHICH CONNECTS THE KIDNEYS TO THE URETERS.	PAIR DUCTS (25 CM) WHICH CONNECT THE KIDNEYS TO THE URINARY BLADDER.
ORGAN WHICH CAN TEMPORARILY STORE UP TO 800ml. OF URINE. CAN CHANGE VOLUME DRAMATICALLY.	A DUCT WHICH CARRIES THE URINE TO THE OUTSIDE OF THE BODY.
THE RELEASE OF URINE FROM THE BLADDER.	THE FUNCTIONAL UNITS OF THE MAMMALIAN KIDNEY. EACH KIDNEY HAS MORE THAN ONE MILLION OF THESE UNITS.
COMPONENT OF NEPHRON, CUPLIKE STRUCTURE	TUBULAR COMPONENT OF THE NEPHRON
CLUSTER OF CAPELLARIES WHICH ARE SITUATED IN BOWNAN'S CAPSULE.	CELLS WHICH MAKE UP THE WALL OF THE GLOMERULUS.

RENAL ARTERIES

COLLECTING DUCT

GLOMERULAR FILTRATE

AFFERENT ARTERIOLES

EFFERENT ARTERIOLE

REABSORPTION

ANTIDIURETIC HORMONE (ADH)

RENAL THRESHOLD

WASTE MATERIALS FROM KIDNEY FILTRATION ARE PASSED INTO THIS TUBE WHICH LEADS TO THE URETER.

DELIVER BLOOD TO THE KIDNEYS

BRANCHES OF THE RENAL ARTERIES WHICH CONDUCT BLOOD INTO THE GLOMERULUS.

THE FLUID WHICH PASSES INTO BOWMAN'S CAPSULE.

THE PROCESS IN WHICH 99% OF THE GLOMERULAR FILTRATE IS RETURNED TO THE BLOOD. ONLY USEFUL SUBSTANCES AND WATER ARE RETURNED.

CARRIES BLOOD FROM THE GLOMERULAR CAPILLARIES TO A SECOND CAPILLARY NETWORK WHICH SURROUNDS THE RENAL TUBULE.

THE MAXIMUM CONCENTRATION OF A SPECIFIC SUBSTANCE IN THE BLOOD AT WHICH COMPLETE RESORPTION CAN TAKE PLACE.

PITUITARY HORMONE WHICH SIGNALS THE WALLS OF THE COLLECTING DUCTS OF THE KIDNEY TO BECOME MORE PERMEABLE TO WATER AND THEREBY INCREASES THE EFFICIENCY OF WATER RESORPTION

CHAPTER 30

ENDOCRINE REGULATION

BEFORE READING ACTIVITIES

1. Before you begin to read the new material, review the important concepts from chapter 29, Disposal of Metabolic Wastes. Use your graphic organizer to give yourself a test. Can you take each of the headings and explain the important concepts? For example, What are waste products? How is the mammalian urinary system designed? Try to put important details into a lower level of your organizer. Use a different color ink when you add details at the lower levels.

2. Turn to page 610. Read the chapter outline about Endocrine Regulation.

   1. What are the three major concepts you need to know about?

      1. 3.

      2.

   2. Why do you think there are separate sections for invertebrate and vertebrate hormones?

   3. Which section is more important, invertebrate hormones or vertebrate hormones? How can you tell?

3. On a separate sheet of paper, construct a graphic organizer to show the relationship between the major and minor ideas. Turn the sheet sideways to leave more room between major ideas.

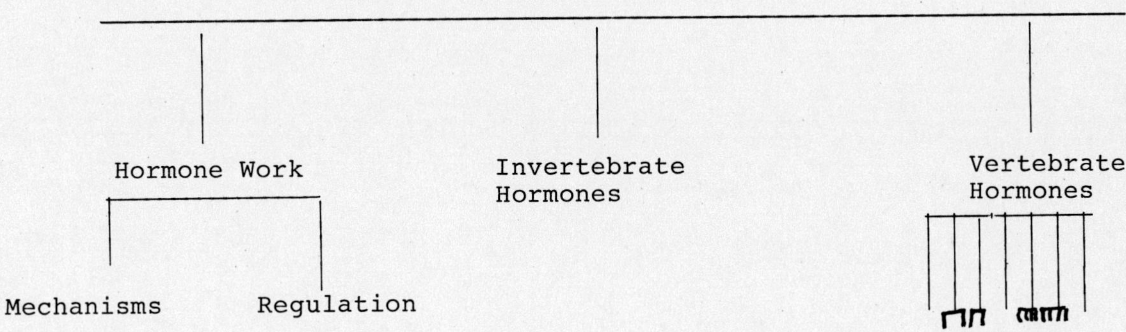

The basic framework of your graphic organizer should look like the one above. You will need to fill in the lower level idea terms from the outline. You should use 3 or 4 different colors of ink to show the level of importance of the ideas.

4. Read the vocabulary cards. There are a lot of new vocabulary terms so you will need to categorize the terms into smaller groups. Some things to look for include the following:

   1. Words that have the same word parts: corpora cardiaca and corpora allata; hyposecretion and hypersecretion; thyroid, thyroxin, and TSH; adrenal glands, adrenal medulla, adrenalin, adrenal cortex; epinephrine and norepinephrine.

   2. Group words together that relate to the same main idea. For example, put all of the terms about the hypothalamus and pituitary gland together.

5. Read the Learning Objectives on page 610. Underline the direction word (key word) in each objective. The words you should have underlined are define, distinguish between, describe, summarize, identify and locate, justify, contrast, give the actions of, and relate.

   1. Go through the textbook chapter quickly and find the pages that contain the information you need to accomplish each objective. Write the page numbers next to the objective.

   2. Think about what you will need to write to accomplish each objective.

DURING READING ACTIVITIES

1. Remember that you need to be able to summarize and describe ideas. Look back at chapter 29 to review what summary and describe mean.

2. In this chapter you are asked to RELATE ideas. Relating ideas means that you must understand how these ideas fit together or CONNECT. For example, how are endocrine glands regulated by negative feedback mechanisms and how is negative feedback related to the hormones discussed in the text?

3. Read the figures in the chapter carefully. Some of the figures may need special attention.

   1. Look at Figure 30-1 on page 611. There are two parts; how is part (a) different from part (b)? What does each part depict? What is the difference between an exocrine gland and an endocrine gland?

   2. Read Figure 30-6 on page 615 carefully. What does it depict?

What are the principal endocrine glands of humans?

1.
2.
3.
4.
5.
6.
7.
8.
9.
10.

Be sure that you can find these glands on a diagram without the labels. Can you locate where these glands are in your own body?

3. Look at Figure 30-9 on page 620. What is the endocrine gland shown?

   What do the arrows mean at the bottom of the diagram?

   What are the 5 important hormones released by the hypothalamus and what does each one cause to happen?

   Prolactin-------- mammary glands----- milk production

   1.
   2.
   3.
   4.
   5.

4. Figure 30-15 on page 629 depicts what happens to YOUR body during times of stress. What are the two major results of stress?

   What does each of these result in?

4. There are two tables in the chapter that summarize a great deal of the information discussed in the chapter.

   1. Table 30-1 lists the principal endocrine glands and their hormones. You may want to reorganize the information into study cards so you can practice test yourself on which glands have which actions.

   2. Table 30-2 on page 618 summarizes the consequences of endocrine malfunction. What will you do to help yourself remember this information?

<u>AFTER READING ACTIVITIES</u>

1. Test yourself on the new vocabulary. If you did not categorize the words appropriately before you read the chapter, regroup the terms that should be in other groups now.

2. Go back to the graphic organizer you made before you began reading this chapter. Explain to yourself (or to someone else) what concepts are involved in each of the headings. Add in the important details at each level, using key terms only, so you will have enough information to help you review this information later. Be sure to color code information by level of importance.

3. Answer the following questions.

   1. What is the major difference between exocrine and endocrine glands?

   2. How does a target tissue recognize a specific hormone?

   3. How do steroid hormones affect cells?

   4. List the principal endocrine glands of the human.

   5. Why is the pituitary gland called "the master gland" of the body?

   6. What hormone is administered to induce labor in a pregnant woman?

   7. Examine Figure 30-9 and answer the following questions.

      A. List the hormones produced by the anterior pituitary.

   8. Excessive hypothyroidism during infancy and childhood can lead to ________________.

   9. Our diet must include _______________ which is necessary for the production of thyroid hormone.

10. ______________________ hormone regulates the levels of calcium in blood and tissue fluids, while __________________ hormone regulates metabolism.

11. The ________________ _______ ___________________ cells in the pancreas produce the hormone insulin which stimulates cells to take up __________________.

12. The two forms of diabetes in humans are ________________________ diabetes and ____________________________ diabetes.

13. Most of the metabolic disturbances associated with diabetes can be traced to three major effects of insulin deficiency. What are these effects?

14. If you are frightened by a car nearly hitting you in a cross walk, the ________________ ______________ gland will have secreted its hormones.

15. Androgens and estrogens are sex hormones which are secreted in _____________ males and females.

Chapter 30 - Answers

1. Exocrine glands release their secretions into ducts, and endocrine glands are ductless.

2. Specific cellular membrane receptor proteins recognize and bind the hormone.

3. Steroid hormones pass through the cell membrane, bind to a cytoplasmic receptor, and this complex moves to the nucleus, where it activates gene functions to produce a particular protein.

4. Pituitary, thymus, ovary/testes, pancreas, adrenals, parathyroids, thyroid, pineal, and hypothalamus.

5. Because through its activity the other endocrine glands are controlled.

6. Oxytocin (pitocin)

7. A. Prolactin; gonadotropic hormones, thyroid stimulating hormone, ACTH, and growth hormone.

8. cretinism

9. iodine

10. Parathyroid, thyroid

11. islets of Langerhans, glucose

12. juvenile-onset, mature-onset

13. Decreased glucose utilization, increased fat metabolism, and increased protein utilization.

14. adrenal medulla

15. both

HORMONES

ENDOCRINE SYSTEM

ENDOCRINE GLANDS

ENDOCRINOLOGY

EXOCRINE GLANDS

TARGET TISSUE

NEGATIVE-FEEDBACK CONTROL MECHANISMS

CYCLIC AMP

HYPOSECRETION

NEUROHORMONES

HYPOTHALAMUS

HYPERSECRETION

DIVERSE COLLECTION OF GLANDS AND TISSUES THAT SECRETE CHEMICAL SUBSTANCES WHICH HELP REGULATE MANY BODILY PROCESSES.

THE CHEMICAL MESSENGER SUBSTANCES SECRETED BY AN ENDOCRINE GLAND. THERE ARE 30-40 DIFFERENT HORMONES: THEY ARE PRESENT IN THE BLOOD IN VERY LOW CONCENTRATION- 1/1,000,000 MILLIGRAMS

THE STUDY OF ENDOCRINE ACTIVITY.

GLANDS WHICH DO NOT HAVE DUCTS THROUGH WHICH THEY RELEASE THEIR HORMONES SECRETIONS. HORMONES ARE RELEASED INTO THE SURROUNDING TISSUE FLUID AND DIFFUSE INTO THE CAPILLARIIES

THE TISSUE OR TISSUES WHICH RESPOND TO HORMONE SIGNALS

GLANDS WHICH RELEASE THEIR SECRETIONS INTO DUCTS.

CYCLIC ADENOSINE MONOPHOSPHATE, THE SECOND MESSENGER, WHICH IS RELEASED IN RESPONSE TO HORMONE STIMULATION. CYCLIC AMP AFFECTS METABOLIC PROCESSES.

INFORMATION REGARDING HORMONE LEVEL IS FED BACK TO THE GLAND WHICH RESPONDS IN A HOMEOSTATIC WAY.

IN INVERTEBRATES HORMONES ARE SECRETED MAINLY BY NEURONS. REGULATE MANY PROCESSES.

ENDOCRINE MALFUNCTION WHERE HORMONE OUTPUT IS DECREASED BELOW NEEDED LEVELS.

ENDOCRINE MALFUNCTION WHERE TOO MUCH HORMONE IS SECRETED.

LINK BETWEEN NERVOUS AND ENDOCRINE SYSTEMS WHICH REGULATES THE ACTIVITY OF THE PITUITARY GLAND.

POSTERIOR LOBE OF THE PITUITARY GLAND

PITUITARY GLAND

PROLACTIN

ANTERIOR LOBE OF THE PITUITARY GLAND

PITUITARY DWARFS

GROWTH HORMONE (SOMATOTROPIN)

THYROID GLAND

GIGANTISM

THYROID STIMULATING HORMONE (TSH)

THYROXINE

GOITER

CRETINISM

GLAND WHICH CONTROLS THE ACTIVITY OF OTHER ENDOCRINE GLANDS: IT IS CALLED THE MASTER GLAND OF THE BODY. SECRETES NINE DISTINCT HORMONES. CONSISTS OF ANTERIOR AND POSTERIOR LOBES.

PRODUCES HORMONES OXYTOCIN AND ANTIDIURETIC HORMONE (ADH).

SECRETES GROWTH HORMONE, PROLACTIN AND SEVERAL TROPIC HORMONES.

HORMONE WHICH STIMULATES MILK PRODUCTION IN THE MAMMARY GLANDS.

HORMONE WHICH STIMULATES BODY GROWTH PRIMARILY BY INCREASING PROTEIN SYNTHESIS.

INDIVIDUALS WHOSE PITUITARY GLAND DID NOT PRODUCE SUFFICIENT GROWTH HORMONE DURING CHILDHOOD.

ABNORMALLY TALL INDIVIDUALS WHOSE ANTERIOR PITUITARY SECRETED EXCESSIVE AMOUNTS OF GROWTH HORMONE DURING CHILDHOOD.

ENDOCRINE GLAND IN THE NECK REGION. SECRETS THYROXINE AND CALCITONIN

IODINE CONTAINING HORMONE WHICH IS ESSENTIAL TO NORMAL GROWTH AND DEVELOPMENT. THIS HORMONE STIMULATES METABOLISM.

REGULATES THE SECRETION OF THYROID HORMONE BY THE THYROID. THIS IS AN ANTERIOR PITUITARY HORMONE. ACTS BY WAY OF CYCLIC AMP.

EXTREME HYPOTHYROIDISM DURING THE EARLY YEARS. RESULTS IN LOW METABOLIC RATE, AND RETARDED MENTAL AND PHYSICAL DEVELOPMENT

AN ABNORMAL ENLARGEMENT OF THE THYROID GLAND.

PARATHYROID HORMONE

PARATHYROID GLANDS

ISLETS OF LANGERHANS OF PANCREAS

CALCITORIM

DIABETES MELLITUS

INSULIN

HYPOGLYCEMIA

HYPERGLYCEMIA

ADRENAL MEDULLA

ADRENAL GLANDS

ANDROGENS

ADRENAL CORTEX

GLANDS EMBEDDED IN CONNECTIVE TISSUE SURROUNDING THE THYROID GLAND. SECRETES PARATHYROID HORMONE.

REGULATES THE CALCIUM LEVEL IN BLOOD AND TISSUE FLUIDS. STIMULATES AN INCREASE IN BLOOD CALCIUM LEVELS

THYROID GLAND HORMONE WHICH WORKS TO REDUCE BLOOD CALCIUM LEVELS.

CLUSTERS OF CELLS IN THE PANCREAS WHICH HAVE THE INDOCRINE FUNCTION OF PRODUCING THE HORMONES GLUCAGON AND INSULIN. THERE ARE ABOUT A MILLION SETS OF THOSE CELLS IN THE PANCREAS.

HORMONE WHICH STIMULATES CELLS TO TAKE UP GLUCOSE. ACTIVITY OF THIS HORMONE LOWERS THE BLOOD GLUCOSE LEVEL.

PRINCIPAL DISORDER OF PANCREATIC HORMONE PRODUCTION

HIGH BLOOD GLUCOSE LEVELS.

LOW BLOOD GLUCOSE LEVELS.

PAIRED GLANDS WHICH ARE SITUATED ON TOP OF THE KIDNEYS. EACH CONSISTS OF AN ADRENAL MEDULLA AND AN ADRENAL CORTEX

CENTRAL PORTION OF THE GLAND. DEVELOPS FROM NEURAL TISSUE. SECRETS EPINEPHRINE AND NOREPINEPHRINE. PREPARES BODY TO DEAL WITH EMERGENCY SITUATIONS

SECRETES STEROID HORMONES: 1) SEX HORMONES, 2) MINERALO-, AND 3) GLUCO CORTICOIDS

HORMONES WHICH HAVE MASCULINIZING EFFECTS.

MINERALOCORTICOIDS

ESTROGENS

CORTISOL

ALDOSTERONE

ACTH

GLUCOCORTICOIDS

PINEAL GLAND

THYMUS GLAND

PROSTAGLANDINS

FEMALE SEX HORMONES.

ADRENAL CORTEX HORMONES WHICH HELP REGULATE SALT BALANCE.

PRINCIPAL MINERALOCORTICOID.

MAIN GLUCOCORTICOID.

ADRENAL CORTICAL HORMONES WHICH PROMOTE GLUCONEOGENESIS IN LIVER

ADRENO CORTICOTROPIC HORMONE FROM PITUITARY. HORMONE WHICH CONTROLS GLUCOCORTICOID SECRETIONS.

RELEASES HORMONE THYMOSIN WHICH PLAYS A ROLE IN IMMUME RESPONSES

PRODUCES MELATONIN WHICH MAY AFFECT REPRODUCTION

HORMONES RELEASED BY MANY DIFFERENT TISSUES WHICH INFLUENCE MANY METABOLIC PROCESSES

CHAPTER 31

REPRODUCTION: PERPEUTATION OF THE SPECIES

BEFORE READING ACTIVITIES

1. Before you begin to read, review the important concepts from chapter 30, Endocrine Regulation. Use the Learning Objectives on page 610 to test yourself. If you cannot remember any information, reread your chapter marking to refresh your memory.

2. Turn to page 632. Read through the outline which gives an overview of the information about reproduction. Answer the following questions.

    1. What are the major concepts discussed?

        1. 4.

        2. 5.

        3.

    2. Which of these concepts is important to you, personally?

    3. What do you expect to be the difference between asexual reproduction and sexual reproduction?

3. On a separate sheet of paper, construct a graphic organizer to show the relationship between the major and minor ideas. Turn the sheet sideways to leave more room between ideas. Be sure to color code the levels of information so you can easily visualize the relationships.

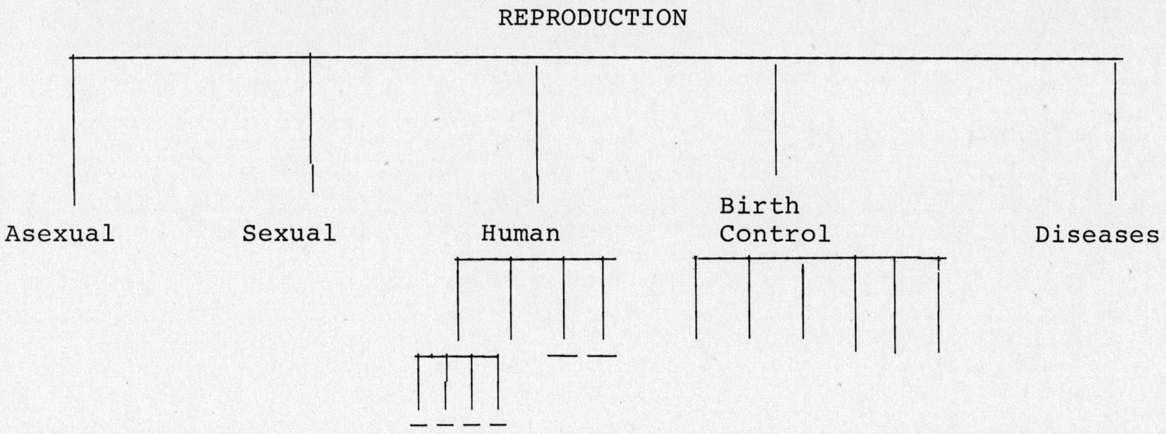

REPRODUCTION

Asexual Sexual Human Birth Control Diseases

The basic structure of your graphic organizer should look like the one above. A look at this organizer should show you that there are two areas that are most important: Human Reproduction and Birth Control. To help you visualize these concepts with all of their contributing components, you should make a separate graphic organizer for each of these areas.

4. Read through the vocabulary cards, the introduction, the summary on pages 659, 660, and 661. Now take the Post-Test on page 661.

5. Tear out the vocabulary cards and categorize the terms; that is, put them into groups that fit together.

6. Read the Learning Objectives on page 632. Underline the direction word (key word) in each objective. The words you should have underlined are compare, trace the development, label on a diagram, give the actions of, describe, identify, summarize, discuss. As you read these, think about what kind of information you need to fulfill each objective. Go through the textbook chapter quickly and find the pages that contain the information you need to accomplish each objective. Write the page numbers next to the objective.

DURING READING ACTIVITIES

1. As you read, think about how you will compare sexual and asexual reproduction. Remember that COMPARE asks you to relate BOTH SIMILARITIES and DIFFERENCES. Be sure you explain (to yourself or to someone else) how these two forms of reproduction are alike as well as how they are different.

2. Two of the objectives ask you to TRACE THE DEVELOPMENT of something. In order to trace the development you must present a SERIES OF RELATED EVENTS. To trace the development and fate of the ovum, you must give each step that occurs in the production of the ovum and each step in the life of the ovum after it is mature.

3. When you are asked to DISCUSS a topic you need to tell WHAT the idea is and WHY. You will have to present the details that explain the idea.

4. As you read this chapter, refine your graphic organizer.

    1. Read through a complete section and then stop to review what you read.

    2. As you review, put the key concept terms in your graphic organizer.

    3. When you read the sections about human reproduction and birth control you will notice that they each have many subsections. This is where you should have a separate graphic organizer to help you visualize the concepts.

5. When you finish reading this chapter, you will be expected to be able to label some diagrams. Be sure you read Figures 31-1, 31-8, and 31-15 carefully. Practice finding and labelling the organs shown on these diagrams. You may want to copy the diagrams without the labels so you can test yourself.

6. Compare Figure 31-7 with Figure 31-16. How are these two figures alike? How are they different? Do both males and females have the same reproductive hormones? Do they have different hormones? Do they produce the same amounts of the same hormones? Tables 31-1 and 31-2 also give information about male and female reproductive hormones. Be sure you read them carefully.

7. There are two other summary tables you should read carefully: Table 31-3 Contraceptive Methods and table 31-5 Some Common Sexually Transmitted Diseases. Each of these tables contains information you need to know.

AFTER READING ACTIVITIES

1. Review your graphic organizer. Explain what each heading means. Be sure you have added important subheadings and color coded them so you will recognize their level of importance. You should also review the graphic organizers you constructed to depict human reproduction and birth control.

2. Review the vocabulary and test yourself. Remember to study the new terms in groups of 5-7 terms at a time.

3. Answer the following questions.

   1. The type of reproduction in which there is not a fusion of gametes is called ______________. Two examples would be ______________ and ____________________. Sexual reproduction requires the union of two different ________________ which when fused is termed a ____________.

   2. Organisms which live in water tend to have ______________ fertilization, whereas those which are terrestrial tend to exhibit ______________ fertilization.

   3. Which parts of the male reproductive system contribute to the non-sperm part of SEMEN?

   4. Sperm are produced in the __________________ _______________ and it takes at least __________ months to produce them. The testes are protected by a sac like structure called the _____________ which keeps the testicles cooler than body temperature.

   5. Trace the movement of sperm through the male reproductive tract.

   6. Identify the organs and organ parts which are uninvolved with a male erection.

   7. Another name for luteinizing hormone (LH) in males is called _____________ _______________ ______________ (ICSH)) which stimulates the ______________ to produce ________________. FSH stimulates the testes to produce ________________.

8. In the female ____________________ __________________ ______________ (FSH) is responsible for stimulating the _______________ to produce ___________. The female hormone _________________ is produced by the corpus luteum and is responsible with estrogen in preparing the __________________ of the uterus for implantation. __________________ _______________ (LH)) is produced in the ____________ _______________ gland and is responsible for causing _____________.

9. Describe the process of menstrual cycle giving some time frame of the process.

10. List the two basic physiologic responses and the four phases of the human sexual response cycle.

11. Refer to Table 31-4; which contraceptive method has the smallest failure rate?

12. Refer to Table 31-5. Which are the most debilitating of the sexually transmitted diseases.

Chapter 31 - Answers

1. asexual, budding, fragmentation, gametes, zygote

2. external, internal

3. prostate gland, seminal vesicle, bulbourethral gland

4. seminiferious tubules, two, scrotum

5. Seminiferous tubules, sertoli cells, epidydimus, vas deferens, ejaculatory duct, seminal vescicles, prostate gland, bulbourethral glands, urethra, end of penis.

6. Nervous system, penis, cavernous bodies, venous senusoids, penal arteries

7. interstitial cell-stimulating hormone, testes, testosterone, sperm

8. follicle stimulating hormone, ovary, ova, progesterone, endometrium, lutenizing hormone, anterior pituitary

9. The menstrual cycle is between 28-32 days normally, the process of ovulation occurs sometime near the middle of the cycle, 14-16 days. FSH starts the process by developing a number of eggs, usually only one comes out of the ovary under the influence of the hormone LH. Estrogens and progesterones are produced in the ovary and help to prepare the endometrium of the uterus for implanting the fertilized egg. If fertilization does not take place the hormones are stopped and menstruation will begin. If fertilization has occurred, the hormones continue to be produced due to human (horionil gonadotropin).

10. Vasocongestion and Muscle Tension. Excitement, plateau, orgasm, resolution.

11. Tubal ligation, douche, oral.

12. Gonorrhea, syphilis, genital herpes, pelvic inflammatory disease

SEXUAL REPRODUCTION

ASEXUAL REPRODUCTION

FRAGMENTATATION

BUDDING

ZYGOTE

GAMETES

FERTILIZATION

HERMAPHRODITIC

EXTERNAL FERTILIZATION

INTERNAL FERTILIZATION

TESTE

GONADS

THE ABILITY TO REPRODUCE WITHOUT MATING SUCH AS BUDDING OR FISSION

REPRODUCTION BY THE FUSSION OF GAMETES, SPERM AND EGG, FORMING A ZYGOTE

A FORM OF ASEXUAL REPRODUCTION IN WHICH A NEW SMALL INDIVIDUAL GROWS OUT OF A PARENT

A FORM OF ASEXUAL REPRODUCTION IN WHICH THE PARENT MAY FRAGMENT INTO MANY PIECES EACH ONE OF WHICH BECOMES A NEW INDIVIDUAL

THE EGG AND SPERM USUALLY DERIVED FROM TWO SEPARATE PARENTS

THE UNION AND FUSSION OF TWO GAMETES, THE PROCESS OF FERTILIZATION

A SINGLE INDIVIDUAL CAPABLE OF PRODUCING BOTH EGG AND SPERM

THE PROCESS WHERE THE NUCLEI OF THE SPERM AND EGG ARE FUSED

REFERRING TO THE DEPOSIT OF THE SPERM INSIDE THE FEMALE

REFERRING TO THE DEPOSIT OF SPERM OUTSIDE OF THE BODY OF THE FEMALE

THE TESTES AND THE OVARIES WHERE THE GAMETES ARE PRODUCED, SPERM AND EGGS

THE MALE ORGAN WHICH PRODUCES SPERM

SPERMATOGONIA

SEMINIFEROUS TUBULES

SERTOLI CELLS

PRIMARY SPERMATOCYTE

SPERMATIDS

SPERMATOGENISIS

INGUINAL CANALS

SCROTUM

STERILE

INGUINAL HERNIA

VAS DEFERENS

EPIDIDYMUS

TINY TUBULES WITHIN THE TESTES IN WHICH SPERM ARE ACTUALLY PRODUCED

THE FIRST OR STEM CELLS FROM WHICH THE SPERM ARE MADE

THE FIRST STAGE IN THE DEVELOPMENT OF SPERM, TO UNDERGO MEIOSIS

CELLS IN THE SEMINIFEROUS TUBULES WHICH NURTURE THE DEVELOPING SPERM

THE PRODUCTION OF SPERM FROM START TO FINISH, A PERIOD OF ABOUT TWO MONTHS

IMMATURE SPERM CELLS

A SKIN COVERED BAG WHICH HOUSES THE TESTICLES

AS THE TESTES DESCEND THE MOVE THROUGH THESE PASSAGEWAYS

OCCASSIONALLY PART OF THE INTESTINE GOES THROUGH THE INGUINAL CANAL AND HERNIATES

THE CONDITION WHERE AN INDIVIDUAL CAN NOT PRODUCE GAMETES

THE LARGER COLLECTING TUBULE WHICH LIES ON TOP OF THE TESTE WHERE SPERM ARE DEVELOPED AND STORED

THE TUBE WHICH LEADS FROM THE EPIDIDYMUS TO THE EJACULATORY DUCT

PROSTATE GLAND

EJACULATORY DUCT

SEMEN

URETHRA

BULBOURETHRAL GLANDS

SEMINAL VESICLES

SHAFT

PENIS

PREPUCE (=FORSKIN)

GLANS

CAVERNOUS BODIES

CIRCUMCISION

THE AREA WHERE SPERM ARE PROPELLED WHEN AN EJACULATION OCCURS

THE GLAND THROUGH WHICH THE EJACULATORY DUCTS PASS AND JOIN WITH THE URETHRA, ALSO PRODUCE SOME OF THE SEMINAL FLUIDS

THE COMMON PASSAGEWAY OF URINE AND SPERM AFTER THE JUNCTION OF THE VAS DEFERENS TO THE URETHRA

THE SPERM AND SECRETIONS AS A TOTAL PART OF THE EJACULATE

PAIRED GLANDS WHICH PRODUCE MANY OF THE FLUIDS OF THE SEMEN

RELEASE SOME ALKALINE SUBSTANCE PRIOR TO EJACULATION TO NEUTRALIZE THE ACID PASSAGE OF URINE WITHIN THE URETHRA

THE ERECTILE COPULATORY ORGAN WHICH DELIVERS SPERM TO THE FEMALE

THE ELONGATE CYLINDRICAL PORTION OF THE PENIS

THE EXPANDED END OF THE PENIS

LOOSE FITTING SKIN OVER THE GLANS PENIS

THE SURGICAL REMOVAL OF THE PREPUCE

THREE PARALLEL CYLINDERS OF ERECTILE TISSUE INSIDE THE PENIS, FILL WITH BLOOD DURING EXCITATION

PUBERTY

ERECTION

LEUTENIZING HORMONE (LH)
(=INTERSTITIAL CELL STIMULATING HORMONE, (ICSH))

FOLLICLE STIMULATING HORMONE (FSH)

MENSTRUAL CYCLE

OVA (=EGGS)

STROMA

OVARY

PRIMARY OOCYTES

OOGONIA

OOGENISIS

FIRST POLAR BODY

THE FILLING OF THE CAVERNOUS BODIES WITH BLOOD TO CAUSE A STIFF ERECT ORGAN

THE PERIOD WHEN YOUNG PERSONS UNDERGO SEXUAL MATURATION

A HORMONE WHICH IN MALES STIMULATES THE SEMINIFEROUS TUBULES TO PRODUCE SPERM, IN FEMALES IT STIMULATES THE OVARIES TO RIPEN AN EGG

A HORMONE WHICH STIMULATES CELLS IN BETWEEN THE SEMINIFEROUS TUBULES TO PRODUCE TESTOSTERONE, IN THE FEMALE IT CAUSES THE EGG TO OVULATE

THE FEMALE GAMETE

THE MONTHLY CYCLE WHICH PREPARES THE OVA FOR POSSIBLE FERTILIZATION

THE FEMALE GONAD WHICH PRODUCES THE EGGS AND THE SEX HORMONES, TESTOSTERONE AND ESTROGEN

THE INFERIOR PORTION OF THE OVARY WHERE THE EGGS DEVELOP

THE CELLS WHICH WILL GIVE RISE TO THE EGGS

THE DEVELOPMENT OF EGG CELLS DURING THE DEVELOPMENT OF THE EMBRYO OF THE FEMALE

AFTER THE FIRST DIVISION ONE CELL DEVELOPS AND THE OTHER IS REDUCED TO A LESSER CELL

THE PROCESS OF DEVELOPING A HAPLOID EGG

THECA

ZONA PELLUCIDA

CORPUS LUTEUM

GRAAFIAN FOLLICLES

TUBAL PREGNANCY

UTERINE TUBE (=FALLOPIAN TUBE)

ENDOMETRIUM

UTERUS

CERVIX

MENSTRUATION

VULVA

PAPANICOLAOU TEST

THE THICK MEMBRANE SURROUNDING THE OOCYTE

AN OUTER COVERING OF CELLS SURROUNDING THE FOLLICLE CELLS

A MATURE FOLLICLE READY TO OVULATE

THE CELLS OF THE FOLLICLE WHICH REMAIN IN THE OVARY AND BEGINS TO SECRETE PROGESTERONE

PAIRED PASSAGEWAYS FROM THE OVARY TO THE UTERUS

A FERTILIZATION WHICH TAKES PLACE AND IMPLANTS IN THE WALL OF THE UTERINE TUBE

THE AREA WHERE IMPLANTATION OF THE EMBRYO WILL TAKE PLACE

THE THICK WALLS OF THE UTERUS WHICH PROVIDES NUTRITION FOR THE RECENTLY IMPLANTED FERTILIZED EGG

THE SLOUGHING OF THE ENDOMETRIAL LINING IF A PREGNANCY HAS NOT OCCURRED

THE LOWER CONSTRICTED PORTION OF THE UTERUS WHICH IS CONTINUOUS WITH THE VAGINA

A TEST TO DETERMINE IF CERVICAL CANCER IS PRESENT

A TERM REFERRING TO THE EXTERNAL FEMALE GENITALIA

LABIA MAJORA

LABIA MINORA

BREASTS

HYMEN

LACTATION

LIGAMENTS

MASTECTOMY

AREOLA

MENAPAUSE

HUMAN CHORIONIC GONADOTROPIN (=HCG)

MUSCLE TENSION

VASOCONGESTION

THE THINNER INNER FOLDS OF SKIN SURROUNDING THE VAGINA

THE THICKER OUTER FOLDS OF SKIN SURROUNDING THE VULVA AND WHICH HAVE HAIR GROWING ON THEIR OUTER SURFACES

A THIN LAYER OF TISSUE WHICH MAY BE PRESENT AND MAY FULLY OR PARTIALLY BLOCK THE OPENING OF THE VAGINA

MODIFIED SWEAT GLANDS WHICH PRODUCE MILK, MAMMARY GLANDS

CONNECT BREASTS TO THE SKIN

THE PRODUCTION AND FLOW OF MILK FROM THE BREAST

THE PINK CENTRAL AREA OF THE BREAST SURROUNDING THE NIPPLE

THE REMOVAL OF A BREAST

THE HORMONE FROM THE PLACENTA WHICH SIGNALS THE CORPUS LUTEUM TO CONTINUE TO PRODUCE PROGESTERONE

THE END OF MENSTRUATION FOR A FEMALE

THE PROCESS OF BLOOD FILLING SPONGY TISSUE AND CAUSING THE TISSUE TO SWELL

THAT WHICH CAUSES THE MUSCLES TO CONTRACT INVOLUNTARILY

PLATEAU PHASE

EJACULATION

ERECTILE DYSFUNCTION (=IMPOTENCE)

ORAL CONTRACEPTIVES

VASECTOMY

SPONTANEOUS ABORTION

SEXUALLY TRANSMITTED DISEASES (=STD)

EXCITEMENT PHASE

ORGASMIC PHASE

RESOLUTION PHASE

CONTRACEPTION

INTERUTERINE DEVICE (=IUD)

TUBAL LIGATION

THERAPUTIC ABORTION

THE STATE OF AROUSING BOTH SEXES

INTENSIFICATION OF EXCITEMENT RAISES TO A LEVEL JUST BEFORE ORGASM

RHYTHMIC CONTRACTIONS OF THE PENIS OR THE VAGINA

THE PROCESS OF EXPELLING SPERM FROM THE END OF THE PENIS

THE RELAXATION AND LETTING GO OF THE TENSION AND VASOCONGESTION

THE INABILITY TO SUSTAIN AN ERECTION

THE PROCESS OF PREVENTING FERTILIZATION

A COMBINATION OF FEMALE HORMONES TO MAKE THE FEMALE BODY THINK ITS PREGNANT

A FOREIGN BODY OF VARIOUS SHAPES WHICH IS PLACED IN THE UTERUS AND THOUGHT TO STOP IMPLANTATION FROM OCCURRING

THE PROCESS OF STERILIZING A MALE BY LIGATING THE VAS DEFERENS

THE PROCESS OF STERILIZING A FEMALE BY LIGATING THE FALLOPIAN TUBES

AN ABORTION WHICH OCCURS BY NATURAL METHODS, WITH NO INTERVENTION

INDUCED TO PRESERVE THE HEALTH OF THE MOTHER

DISEASES WHICH EFFECT THE REPRODUCTIVE SYSTEM AND HAVE WIDENING EFFECTS ON THE ORGANISM

CHAPTER 32

DEVELOPMENT: THE ORIGIN OF THE ORGANISM

BEFORE READING ACTIVITIES

1. Before you begin to read about the origin of the organism, review the information about reproduction in chapter 31. Remember that these two chapters contain many concepts in common; chapter 32 is a continuation of the process discussed in chapter 31.

2. Turn to page 663 and read the outline about development. Answer the following questions.
   1. What are the major topics to be discussed?

      1. 　　　　　　4.

      2. 　　　　　　5.

      3. 　　　　　　6.

   2. What do you already know about the pattern of development? birth? adjusting to life after birth? environmental influences on the embryo? the life cycle? the aging process?

   3. Which of these topics do you know the least about? (You will need to study these most.)

3. Read through the vocabulary cards. You are probably already familiar with many of the terms included; be sure that you know the biological definition of each term.

4. Read the introduction on page 644 (carefully!), the summary on pages 680 and 681 and take the Post-Test on page 681.

5. Read the Learning Objectives on page 663. Underline the direction word in each objective. Go through the textbook chapter quickly and find the pages that contain the information you need to know to accomplish each objective. Write the page numbers next to the objective.

DURING READING ACTIVITIES

1. Be sure you underline or highlight as you read. After you have done this, summarize the information in your own words in the margin. Remember to read a whole paragraph or group of paragraphs before you make any marks. The example below is from page 678 of the textbook.

   THE AGING PROCESS

   Since development in its broadest sense includes any biological change with time, it also includes those changes that result in the decreased functional capacities of the mature organism, the changes commonly called aging. The declining capacities of the various systems in the human body, though most apparent in the elderly, may begin much earlier in life, during childhood, or even during prenatal life. The newborn female has only 400,000 oocytes remaining of the 4,000,000 she had three months earlier in fetal life.

The aging process is far from uniform in various parts of the body. Various systems of the body may begin their decline at quite different times. A 75-year old man, for example, has lost 64% of his taste buds, 44% of the renal glomeruli, and 37% of the axons in his spinal nerves that he had at age 30. His nerve impulses are propagated at a rate 10% slower, the blood supply to his brain is 20% less, and the vital capacity of his lungs has declined 44%. The aging process is also marked by a progressive decrease in the body's homoestatic ability to respond to stress.

1. Check your own marking to see if it is similar to this.

2. Remember that you should not mark whole sentences. Mark only the important words in a sentence.

3. Did you circle the vocabulary term (aging) and underline the definition?

4. Did you mark the example of non-uniformity of the aging process?

2. Read tables 32-2, 32-3, and 32-4 carefully. You may need to reorganize this information in some way so you will be able to remember it when you need it.

AFTER READING ACTIVITIES

1. Review your textbook marking carefully. If your textbook marking is accurate, you should not have to reread your whole text when you review for an examination.

2. Review your vocabulary. Test yourself by writing the definitions or by saying them aloud. It is important that you review these terms in an active way, writing or speaking, and not just by reading them silently.

3. Answer the questions which follow:

1. Development of an organism involves three coordinated processes, these are ____________________, ________________________ and ______________________________.

2. How is the human (mammalian) embryo nourished during the first few days?

3. What happens when the inner cell mass divides into two masses?

4. Distinguish between the origin of identical twins and fraternal twins.

5. Examine Figure 32-2 and answer the following questions.

   A. By examining diagrams (a) and (b), what has occurred between days 7 and 10?

   B. The amniotic and chorionic cavities surround the embryo. Which one is closest to the embryo?

6. What tissues contribute to the formation of the placenta?

7. The placenta brings the maternal blood close to the blood of the embryo, but the two circulatory systems are ____________________ ____________________.

8. Human chorionic gonadotrophin stimulates the corpus luteum to produce ____________________ and ____________________. These two hormones in turn cause continued development of the uterine _______________ and ____________________.

9. The three primary germ layers are the _______________, _______________, and the ______________________.

10. Examine Table 32-1 and answer the following questions.

    A. Name the three primary germ layers from the outside to the inside of the embryo.

B. List the two main sets of structures formed by the ectoderm.

C. List the major structures formed by the endoderm.

List the structures formed by the mesoderm.

11. How long after fertilization do the major organ systems form?

12. What is the average gestation period for a human?

13. What is another name for the "bag of waters"?

14. Why is it better for a woman to have natural childbirth?

15. In the United States approximately ____________ percent of all live births exhibit a significant clinical defect. These defects may result from _________________ and/or ____________________ factors.

16. If a mother consumes too much alcohol during her pregnancy, the child may have ________________ ________________ ______________. This may lead to a deformed, physically and mentally retarded child.

17. Cigarette smoking may result in _______________ babies.

18. Mothers who use drugs during their pregnancy increase the risk of high ______________ and prematurity rates.

Chapter 32 - Answers

1. growth, morphogenesis, cell differentiation

2. Embryo is bathed in nutritive fluid secreted by glands of the uterus.

3. Identical twins form.

4. Identical twins are formed from subdivisions of a genetically identical inner cell mass. Fraternal twins form when two distinct ova are fertilized.

5. A. The developing embryo has invaded the uterine wall, and is now embedded in the uterine wall.

   B. The amniotic cavity.

6. The embryonic chorion and the maternal uterine tissue.

7. completely separate

8. progesterone, estrogen, endometrium, placenta

9. ectoderm, mesoderm, endoderm

10. A. Ectoderm, mesoderm, endoderm.

    B. The nervous system and outer layer of skin (epidermis).

    C. Lining of the digestive tube

Skeleton, all muscles, circulatory system, excretory system, reproductive system, dermis, muscle layers, of the digestive tube.

11. approximately four weeks.

12. 266 days.

13. Amniotic cavity.

14. Because the anesthetics affect the newborn infant and may affect breathing

15. seven, genetic, environmental

16. fetal alcohol syndrome

17. smaller

18. mortality

THEORY OF EPIGENESIS

PREFORMATION THEORY

MORPHOGENESIS

GROWTH

ZYGOTE

CELLULAR DIFFERENTIATION

MORULA

CLEAVAGE STAGE

TROPHOBLAST

BLASTOCYST (BLASTULA)

CONJOINED TWINS

INNER CELL MASS

THEORY THAT BELIEVED THE EGG OR SPERM CONTAINED A COMPLETELY FORMED MINIATURE BEING.

THEORY THAT THE EMBRYO DEVELOPS FROM AN UNDIFFERENTIATED ZYGOTE AND THAT THE STRUCTURES OF THE BODY EMERGE IN AN ORDERLY SEQUENCE.

IN THE CONTEXT OF EMBRYOS REFERS TO BOTH CELLULAR GROWTH AND MITOSIS.

THE PRECISE, COMPLICATED CELLULAR MOVEMENTS THAT BRING ABOUT THE FORM OF A MULTICELLULAR ORGANISM WITH ITS PATTERN OF TISSUES AND ORGANS.

THE PROCESS BY WHICH CELLS BECAME SPECIALIZED.

THE NEWLY FERTILIZED EGG.

SERIES OF MITOSES WHICH OCCUR IN THE EMBRYO SHORTLY AFTER FERTILIZATION.

STAGE OF DEVELOPMENT WHEN THE EMBRYO CONSISTS OF A CLUSTER OF CELLS (16-CELL STAGE IN HUMANS).

EARLY DEVELOPMENTAL STAGE WHERE EMBRYO CELLS FORM A HOLLOW BALL.

OUTER LAYER OF BLASTOCYST CELLS WHICH WILL FORM NON-EMBRYO STRUCTURES, THE PROTECTIVE AND NUTRITIVE MEMBRANES WHICH SURROUND THE EMBRYO.

A CENTRAL CLUSTER OF BLASTOCYST CELLS WHICH ACTUALLY GIVE RISE TO THE EMBRYO ITSELF.

TWINS FORMED WHEN THERE IS INCOMPLETE SEPARATION OF THE INNER CELL MASS; THESE TWINS ARE PHYSICALLY JOINED AT SOME BODY PART.

IMPLANT

EXTRAEMBRYONIC MEMBRANES

YOLK SAC

FETAL MEMBRANES

ALLANTOIS

CHORION

PLACENTA

AMNION

EMBRYONIC DISC

HUMAN CHORIONIC GONADOTROPHIN (HCG)

ECTODERM

ENDODERM

MEMBRANES WHICH ARE NOT PART OF THE EMBRYO ITSELF, AND WHICH PROTECT THE EMBRYO, AND HELP IT OBTAIN FOOD AND OXYGEN AND ELIMINATE WASTE.

THE PROCESS IN WHICH THE BLASTOCYST EMBEDS ITSELF IN THE UTERINE WALL.

NONEMBRYONIC STRUCTURES WHICH INCLUDE THE AMNION, THE YOLK SAC, THE CHORION AND THE ALLANTOIS.

FORMS AS AN OUTPOCKETING OF THE DEVELOPING OF GUT. TEMPORARILY INVOLVED IN RED BLOOD CELL FORMATION

NONEMBRYO STRUCTURE WHICH DEVELOPS FROM TROPHOBLAST. SURROUNDS DEVELOPING EMBRYO. APPEARS TO BE VESTIGIAL.

OUTPOCKETING FROM PRIMITIVE GUT. APPEARS TO BE VESTIGIAL ORGAN.

A THIN MEMBRANE WHICH FORMS EARLY IN DEVELOPMENT AND SURROUNDS THE DEVELOPING EMBRYO. THIS MEMBRANE FORMS THE FLUID-FILLED AMNIOTIC CAVITY. THE AMNIOTIC FLUID (IN THE AMNIOTIC CAVITY) BATHES THE EMBRYO, AND SERVES TO PROTECT IT.

IN PLACENTAL MAMMALS THIS STRUCTURE IS THE ORGAN WHICH ALLOWS EXCHANGE OF NUTRIENTS, OXYGEN AND WASTES BETWEEN THE MOTHERS BLOOD SUPPLY AND THAT OF THE EMBRYO. ALSO, SERVICES AS AN ENDOCRINE ORGAN.

A HORMONE SECRETED BY THE TROPHOBLAST CELLS WHICH SIGNAL THE CORPUS LUTEUM IN THE OVARY TO PRODUCE PROGESTERONE AND ESTROGEN.

THE CELLS OF THE INNER CELL MASS ARE ARRANGED INTO TWO LAYERS AND THEN THREE LAYERS FROM WHICH THE EMBRYO FORMS.

THE CELLS OF THE LOWER LEVEL OF THE EMBRYONIC DISCH WHICH WILL GIVE RISE TO THE LINING OF THE DIGESTIVE TRACT.

THE OUTERMOST LAYER OF THE EMBRYO WHICH WILL EVENTUALLY COVER THE ENTIRE EMBRYO AND GIVE RISE TO THE SKIN AND NEURAL STRUCTURES.

NEURAL PLATE

GERM LAYERS

NOTOCHORD

GASTRULATION

FETUS

NEURAL TUBE

LABOR

LANUGO

NEONATE

EPISIOTOMY

SONOGRAM

AMNIOCENTESIS

THE THREE LAYERS OF AN EARLY EMBRYO--ECTODERM, MESODERM, AND ENDODERM. EACH OF THESE THREE LAYERS GIVES RISE TO SPECIFIC STRUCTURES.

THICKENED PLATE OF ECTODERM WHICH FORMS AT GASTRULATION, AND GIVES RISE TO THE NEURAL STRUCTURES.

THE PROCESS BY WHICH THE BLASTOCYST DEVELOPS INTO A THREE-LAYERED STRUCTURE, THE GASTRULA.

A FLEXIBLE ROD OF TISSUE WHICH FORMS BELOW THE NEURAL TUBE. SERVES AS SKELETAL AXIS IN CHORDATE EMBRYOS. NOTOCHORD IS REPLACED BY THE VERTEBRAL COLUMN IN VERTEBRATE EMBRYOS.

CYLINDRICAL STRUCTURE FORMED BY ROUNDING-UP OF THE NEURAL PLATE. DIFFERENTIATES INTO BRAIN AND ALL NERVOUS SYSTEM STRUCTURES.

HUMAN EMBRYO AFTER TWO MONTHS OF DEVELOPMENT. BY THIS TIME BASIC FORM OF ALL ORGANS HAVE BEEN LAID DOWN.

LAYER OF DOWNY HAIR COVERING THE FETUS WHICH IS USUALLY SHED AT BIRTH.

SERIES OF INVOLUNTARY CONTRACTIONS OF THE UTERUS.

AN INCISION MADE BY THE PHYSICIAN DURING CHILDBIRTH. THIS INCISION EXTENDS FROM THE VAGINA TOWARDS THE ANUS AND PREVENTS TEARING OF THE MATERNAL TISSUES.

NEWBORN INFANT FROM BIRTH TO THE END OF THE FIRST MONTH.

TECHNIQUE IN WHICH AMNIOTIC FLUID IS REMOVED, AND THE CELLS AND FLUID CONTAINED THEREIN ARE TESTED FOR SEVERAL GENETIC AND BIOCHEMICAL DEFECTS.

ULTRASOUND PHOTOGRAPH TAKEN OF EMBRYO. USED TO DETERMINE AGE OF THE EMBRYO, AND TO DIAGNOSE DEFECTS.

CHILDHOOD

INFANCY

AGING

ADOLESCENCE

IN VITRO FERTILIZATION

ARTIFICIAL INSEMINATION

FRATERNAL TWINS

FOLLOWS THE NEONATAL PERIOD AND LASTS UNTIL THE INFANT CAN ASSUME AN ERECT POSTURE AND WALK (10-14 MONTHS OF AGE).

PERIOD OF RAPID GROWTH AND DEVELOPMENT WHICH LASTS FROM INFANCY TO ADOLESCENCE.

THE TIME OF DEVELOPMENT BETWEEN PUBERTY AND ADULTHOOD.

CHANGES THAT RESULT IN THE DECREASED FUNCTIONAL CAPACITIES OF THE MATURE ORGANISM.

PROCEDURE IN WHICH DONOR SPERM ARE USED TO FERTILIZE A WOMAN'S EGG. USED WHEN MALE IS STERILE OR HAS GENETIC DEFECT(S).

TECHNIQUE IN WHICH A WOMAN'S EGG IS REMOVED, FERTILIZED IN A TEST TUBE AND REPLACED IN THE WOMAN'S UTERUS.

TWINS FORMED WHEN TWO OVA ARE RELEASED, FERTILIZED, AND DEVELOP INTO TWO COMPLETE ORGANISMS.

CHAPTER 33

VARIABILITY AND EVOLUTION

BEFORE READING ACTIVITY

1. This is the first chapter in a new section of the text with the title "Evolution, Behavior, and Ecology." Each chapter in this section is related to the other chapters in the section and discusses some aspect of evolution behavior, and ecology. Although you will read them as separate units, it is important that you view them as parts of a whole unit.

2. Turn to the outline on page 685. Read it and answer the following questions.

   1. How many major concepts are there in this chapter? What are they?

      1.

      2.

      3.

      4.

      5.

   2. What are the most important major ideas discussed and how can you tell?

      1.

      2.

   3. What terms are used in the outline that you already know? Are there terms that you know as common language that may have different meanings in this context (such as immigration and emigration)?

3. Read through the Learning Objectives on page 685. Notice that the key word used most often is "discuss." Recall that discuss means that you need to be able to tell what an idea is, present the details that explain the idea and tell why the idea is distinctive and/or related to something.

4. Read through the vocabulary items. Some things to notice include:

   1. There are many terms you have learned in previous chapters, such as heterozygote, allelic genes.

   2. There are word parts that you should be familiar with, such as micro- and macro-.

   3. There are some terms which have common meanings, but have different meanings in this context, such as founder, bottleneck, mimicry.

4. Read the introduction, summary, review the vocabulary, and take the Post-Test on page 701.

DURING READING ACTIVITIES

1. Read the chapter in sections. Mark the paragraphs in each section AFTER you read them through completely. Use your own marking system to show what the main ideas are. You may want to underline main ideas twice, asterisk them, or use a wavy line to show main ideas.

2. Be sure to use SMALL, CIRCLED NUMBERS next to words which are part of a group of items about the same idea. For example, the points of the Darwin and Wallace proposal given on page 688 should be numbered. You should also write these points in the margin to help you remember them and find them easily.

3. Notice as you read this chapter that the writing style is different from the writing style in previous chapters. In the earlier chapters there were many facts packed closely together in each paragraph. In this chapter you will find more general ideas and less specific details in each paragraph. You still need to find the main ideas and supporting details. Since much of the information in this chapter is a discussion of theories, you need to find the following information about each theory:

    1. Who proposed the theory.

    2. What is the theory.

    3. Why did the theorist suggest this theory?

    4. What scientific evidence is there to support this theory?

    5. Is this theory still used? Why or why not? If not, what has taken its place.

AFTER READING ACTIVITIES

1. Review all of your margin notes and other markings. Be sure that you understand what you marked and that you did not omit important information. Be careful not to mark too much information. If you find some irrelevant information marked, remove your marks in some way.

2. Review the vocabulary. Practice the terms by writing them or saying them aloud.

3. Answer the following questions.

    1. The term ____________________ ______________ refers to change through time of living oganisms. This can occur at the SPECIES level in which case it is called ________________________ or if it occurs at a level higher than SPECIES it is termed _________________.

    2. List the major contributors to the early thoughts about EVOLUTION and write a brief statement about their contributions.

3. If a population is allowed to reproduce unchecked by natural selection, mutation, immigration, emigration, or other such influences the ____________ ________________ will ___________ ____________ ____________.

4. List those phenomena which will change gene frequencies.

5. How does PROTECTIVE COLORATION demonstrate NATURAL SELECTION?

6. The type of SPECIATION where two species occur in the same area is called ______________. An example of how this could happen would be the doubling of chromosomes which is termed _______________. Another type of speciation is called _____________, in which a large population becomes geographically isolated; then natural selection works on either population or both and they become reproductively isolated thereby becoming ________ __________.

7. Methods which help to maintain species genetic isolation are _____________ ____________ ___________ which are typified by things like behavioral differences. Another type of isolation is termed __________________ _____________ ______________, an example of which could be hybrid sterility.

Chapter 33 - Answers

1. organic evolution, microevolution, macroevolution

2. CHARLES DARWIN published the first works on EVOLUTION which stated four main points: (1) overproduction of offspring, (2) variation in populations, (3) variation is inherited, (4) natural selection of those suited to their environment.
   LAMARCK thought that characteristics were acquired by use and disuse.
   WALLACE came to similar conclusions as DARWIN at the same time.

3. gene frequencies, remain the same

4. List those phenomena which will change gene frequencies.
   (1) mutation, (2) natural selection, (3) immigration, (4) emmigration, (5) genetic drift, (6) founder effect, (7) bottleneck effect, (8) incomplete interbreeding.

5. NATURAL SELECTION is the process whereby organisms which are well suited to their environment are selected for whereas those who are not are selected against. If an organism is protected by its environment in some way then those which best match the environment will be selected for.)

6. sympatric, polyploidy, allopatric, two species

7. prezygotic isolating mechanisms, postzygotic isolating mechanism

ORGANIC EVOLUTION

EVOLUTION

MACROEVOLUTION

MICROEVOLUTION

LAMARCK

CHARLES DARWIN

ALFRED WALLACE

ON THE ORIGIN OF SPECIES BY MEANS OF NATURAL SELECTION

PANMICTIC

NEO-DARWINISM

HARDY-WEINBERG LAW

POOL OF ALLELIC GENES

CHANGE THROUGH TIME	CHANGE THROUGH TIME OF LIVING THINGS
CHANGES IN BREEDING POPULATIONS, USUALLY SPECIES, WHEN GENE FREQUENCIES ARE DISTURBED WITHIN THE BREEDING UNIT.	CHANGES AT HIGHER LEVELS, ABOVE THE SPECIES LEVEL, GIVING RISE TO NEW GROUPS AT THE HIGHER LEVEL
THE FATHER OF MODERN EVOLUTIONARY THINKING	PROPOSED A THEORY OF EVOLUTION BASED ON THE USE AND DISUSE OF ORGANS, E.G., GIRAFFES STRETCHING THEIR NECKS TO GET FOOD
THE TITLE OF THE BOOK WHICH DARWIN WROTE REGARDING HIS THEORY OF EVOLUTION IN 1859	A FELLOW NATURALIST WHO CAME TO SIMILAR IDEAS REGARDING EVOLUTION INDEPENDENT OF DARWIN AT THE SAME TIME
THE EXTENTION OF DARWIN'S IDEAS BY INCLUDING INFORMATION ABOUT GENETICS	AN INTERBREEDING POPULATION IN WHICH ALL GENES WITHIN THE POPULATION CAN COMBINE WITH ANY OTHER GENE
THE GENES, AND ALL THEIR VARIATION, WHICH ARE IN THE BREEDING UNIT	THE LAW WHICH STATES THE IF A POPULATION IS ALLOWED TO CONTINUE BREEDING WITHOUT BEING DISRUPTED BY A VARIETY OF FORCES THAT THE ALLELIC GENES WILL REMAIN IN THE SAME FREQUENCY INDEFINITELY.

POLYPLOIDS

MUTATION

IMMIGRATION

NATURAL SELECTION

EMIGRATION

GENETIC DRIFT

BOTTLENECK EFFECT

FOUNDER EFFECT

MIMICRY

PROTECTIVE COLORATION

GENETIC RECOMBINATION

BATESIAN MIMICRY

ANY CHANGE IN THE GENE WHICH IN TURN AFFECTS THE EXPRESSION OF THAT GENE, THE ULTIMATE RAW MATERIAL FOR EVOLUTIONARY CHANGES

THE PROCESS OF MULTIPLYING, USUALLY DOUBLING, THE CHROMOSOME NUMBER OF AN INDIVIDUAL BY SOME DEFECTS DURING MEISOIS THEREBY CAUSING VARIABILITY TO OCCUR

THE PROCESS OF SELECTING THE ORGANISM, AND THEREFORE ITS GENES, WHICH IS BEST SUITED TO ITS ENVIRONMENT AND THEREFORE PRESERVING ITS GENES FOR FUTURE GENERATIONS.

THE MOVEMENT OF AN INDIVIDUAL OR INDIVIDUALS INTO A NEW POPULATION THEREBY INTRODUCING NEW OR MORE GENES THUS CHANGING THE FREQUENCY OF THOSE GENES

THE MOVEMENT OF AN INDIVIDUAL OR INDIVIDUALS OUT OF A POPULATION THEREBY REMOVING GENES FROM THE BREEDING UNIT THUS CHANGING THE POOL OF ALLELIC GENES

THE CHANGES OF GENE FREQUENCIES BY RANDOM EVENTS

WHEN A NEW POPULATION ARRIVES AT A NEW HABITAT THE POOL OF ALLELIC GENES IS RESTRICTED TO THE FOUNDERS OF THE NEW POPULATIONS AND THEREFORE DOES NOT CARRY THE SAME GENETIC INFORMATION AS THE ORIGINAL POPULATION

THE PERIODIC ELIMINATION OF CERTAIN INDIVIDUALS WHICH LEAVES THE BREEDING POPULATION RESTRICTED TO A DIMINISHED POOL OF ALLELIC GENES

A CERTAIN PATTERN OF COLOR OF AN ORGANISM WHICH IN SOME WAY GIVES IT AN ADVANTAGE LIVING IN THE HABITAT WHERE IT LIVES

A SYSTEM WHEREBY ONE SPECIES MIMICS ANOTHER SPECIES CALLED THE MODEL

A SPECIAL KIND OF MIMICRY WHERE THE MODEL IS SOMEHOW DANGEROUS OR DISTASTEFUL AND THE MIMIC ONLY LOOKS LIKE THE MODEL.

THE ABILITY OF AN INDIVIDUAL TO TRANSMIT ITS GENES TO FUTURE GENERATIONS

SPECIATION

HYBRID VIGOR

ALLOPATRIC SPECIATION

SPECIES

ISOLATING MECHANISMS

SYMPATRIC SPECIATION

POSTZYGOTIC ISOLATION

PREZYGOTIC ISOLATION

MACROMUTATION

STABILIZING SELECTION

SALTATION

REGULATORY GENES

PUNCTUATED EVOLUTION

WHEN NEW COMBINATIONS OF GENES PRODUCE INDIVIDUALS WHICH ARE MORE FAVORABLE THAN THE ORIGINAL POPULATION

THE PROCESS OF REPRODUCTIVELY ISOLATING TWO POPULATIONS FROM ONE ANOTHER

A GROUP OF REPRODUCTIVELY, THEREFORE, GENETICALLY, ISOLATED INDIVIDUALS

THE PROCESS OF DEVELOPING A NEW SPECIES BY ISOLATING AT LEAST TWO PORTIONS OF A FORMERLY SINGLE BREEDING (SPECIES) AND THROUGH TIME AND NATURAL SELECTION PRODUCE TWO REPRODUCTIVELY ISOLATED POPULATIONS

TWO POPULATIONS OCCUPYING THE SAME TERRITORY, PERHAPS THROUGH GENE DIFFERENTIATION OR POLYPLOIDY CREATE REPRODUCTIVELY ISOLATED POPULATIONS

VARIOUS MECHANISMS WHICH ASSURE THAT TWO SPECIES WILL MAINTAIN THEIR REPRODUCTIVE ISOLATION

MECHANISMS WHICH ISOLATE TWO SPECIES BEFORE A ZYGOTE IS FORMED

MECHANISMS WHICH ISOLATE TWO SPECIES AFTER A ZYGOTE IS FORMED

THE MAINTENANCE OF A RANGE OF VARIATION BUT TO SELECT A STANDARD PHENOTYPE WITHIN THE POPULATION

MAJOR MUTATIONS WHICH AFFECT LARGE NUMBERS OF SPECIES AT THE SAME TIME

THOSE GENES WHICH ACCOMPLISH MAJOR REGULATORY ACTIVITY SUCH AS GROWTH RATE

SIMPLE CHANGES IN REGULATORY GENES WHICH MIGHT PRODUCE LARGE EFFECTS.

PERIODS OF NO EVOLUTIONARY ACTIVITY FOLLOWED BY MAJOR (PUNCTUATED) PERIODS OF SPECIATION.

CHAPTER 34

EVOLUTION: ORIGINS AND EVIDENCE

BEFORE READING ACTIVITIES

1. Review the important ideas presented in Chapter 33. Notice that both chapter 33 and chapter 34 discuss aspects of evolution. You will need to keep in mind what you learned in chapter 33 to understand chapter 34.

2. You may want to review the information in the first part of the text, the Organization of Life, before you begin to read this chapter. The information in chapters 2 and 4 will help you to understand chapter 34 more completely.

3. Turn to the outline on page 702. Read it and answer the questions which follow.

   1. What are the concepts to be discussed related to the origin of life?

      1. 3.

      2. 4.

   2. What do you think is meant by the term, "evolutionary evidence?"

   3. What are the types of evolutionary evidence discussed in this chapter:

      1. 5.

      2. 6.

      3. 7.

      4. 8.

4. Read through the Learning Objectives on page 702. Notice that in objective 3 you are asked to "critically discuss" types of evidence. CRITICALLY DISCUSS means that you must EXAMINE, ANALYZE, and EVALUATE the EVIDENCE presented in the text. Questions like this are the most difficult type to answer because you must first summarize the information and then discuss the merits of the information.

5. Read through the vocabulary cards. Notice that there are not many new terms.

6. Read the introduction to this chapter <u>carefully</u>. This introduction sets the tone for the rest of the chapter. Read the chapter summary on page 722. Take the Post-Test.

DURING READING ACTIVITIES

1. Continue to read the chapter in sections, from one heading to the next. Be sure to mark the important points and supporting details. Do not mark too much. If you mark too much information, you will have to reread unimportant material when you review.

2. Be aware of the qualifying words used in this chapter: according to this view, two ways have been proposed, would probably support, principal evidence, and many others. The authors are clearly trying to present information so that you can develop an opinion based on th evidence presented.

3. As suggested in the previous chapter, you need to have answers for several questions about each theory.

AFTER READING ACTIVITIES

1. Check all of your textbook marking. This is a good way to review the chapter. Each time you review, you strengthen the memory traces in your brain; the stronger memory traces are more likely to lead to the ability to recall information when you need it.

2. Review the vocabulary from this chapter and from chapter 33.

3. Answer the following questions.

   1. Briefly describe the experiments which best explain the origin of ORGANIC MOLECULES.

   2. Briefly describe how cells could have formed.

   3. The ____________________ ____________ suggests that cellular __________ originally were independent and then became engulfed into some cell membrane to become a _____________ cell with a nucleus and organelles.

   4. A possible precursor to th EUKARYOTIC cell may have been an anerobic form called ___________________. Another old rock formation called _______________ carry some bluegreen algae thought to be the oldest fossils in the world.

   5. The study of the evolution of humans started with a discovery of _____________ _________. This discovery was followed by _________ (____________) which lead to modern man. Another branch in the HOMINID line lead to the discovery of _____________________ which is now extinct.

   6. List the six types of evidence which are used to support ORGANIC EVOLUTION.

7. A structure like the wing of an insect when compared to that of a bird is an ________________ ______________ whereas if one compares the wing of a bat to one of a bird it is an ________________ ___________________.

8. ________________ ______________ is the evolutionary process by which an ancestral species gives rise to many diverse forms.

9. Identify two ways that rocks and fossils can be aged.

Chapter 34 - Answers

1. If scientists put a mixture of nitrogen, carbon dioxide, methane, ammonia, hydrogen, and water vapor and place them in some condition exposing them to a high energy source they will produce some complex organic molecules.

2. Nucleic acids under the influence of catalysts could duplicate themselves and produce enzymes, the early nucleic acids may have been RNA followed by DNA.

3. endosymbiotic theory, organelles, eukaryotic

4. archaebacteria, stromatolites

5. neandrathal man, australopithetcus

6. (1) microevolution, (2) comparative anatomy, (3) biochemistry, (4) genetics, (5) distribution, (6) fossil record

7. analogous character, homologous character

8. adaptive radiation

9. Potassium-Argon Method and Radiocarbon Dating.

ORGANIC MOLECULES

ABIOGENESIS

PROKARYOTE

COACERVATES

ENDOSYMBIOTIC THEORY

EUKARYOTE

LAW OF SUPERPOSITION

ARCHAEBACTERIA

STROMATOLITES

SUPRAORBITAL TORUS

HOMINIDS

NEANDERTHAL MAN

THE THEORY THAT LIFE WAS GENERATED FROM NONLIVING MOLECULES.

CHEMICAL COMPOUNDS WHICH CONTAIN CARBON.

MACROMOLECULES WHICH COLLECT TOGETHER AND ACT AS CATALYSTS, SHARE PROPERTIES OF LIVING CELLS.

UNICELLULAR ORGANISMS, THE SIMPLEST OF ALL CELLS WHICH DO NOT HAVE A NUCLEAR MEMBRANE OR ORGANELLES.

UNICELLULAR ORGANISMS WHICH HAVE A NUCLEAR MEMBRANE AND ORGANELLES.

THE THEORY WHICH SUGGESTS THAT ORGANELLES WHICH POSSESS MEMBRANES MAY HAVE BEEN INTRODUCED INTO PROKARYOTE TO MAKE EUKARYOTES.

A POTENTIAL ANCESTOR OF EUKARYOTIC CELLS, ANEROBIC FORMS.

THE LAW WHICH STATES THAT YOUNGER ROCKS LIE ON TOP OF OLDER ROCKS, THEREFORE A CHRONOLOGY OF FOSSILS CAN BE CREATED.

BLUE GREEN ALGAE AGGREGATIONS FOUND IN CERTAIN OF THE OLDEST ROCKS IN THE EARTH.

THE HEAVY RIDGE OF BONE OVER THE EYES OF APE SKULLS, USED IN DETERMINING RELATIONSHIPS AMONG PRIMATES.

THE FIRST DISCOVERY OF A FOSSIL HUMAN.

THE GROUP OF HUMAN AND HUMAN-LIKE ORGANISMS.

AUSTRALOPITHECUS

HOMO ERECTUS

ANALOGOUS CHARACTERS

HOMOLOGOUS CHARACTERS

BIOGEOGRAPHY

VESTIGAL CHARACTERS

GAUSE'S LAW

CENTER OF ORIGIN

NICHE

STRUCTURAL PARALLELISM

DIVERGENT EVOLUTION

ADAPTIVE RADIATION

RADIOCARBON DATING

POTASSIUM-ARGON METHOD

THE SCIENTIFIC NAME GIVEN TO THE FIRST RECOGNIZABLE HUMAN SKELETON WHICH INDICATED THAT THE INDIVIDUAL WALKED IN AN UPRIGHT POSITION, SMALLER THAN HOMO SAPIENS AND WITH A SMALLER BRAIN CASE.

A BRANCH IN HOMINID EVOLUTION WHICH BECAME EXTINCT.

CHARACTERS WHICH CAN BE SHOWN TO HAVE A SIMILAR ANATOMY AND A SIMILAR ORIGIN.

CHARACTERS WHICH HAVE SIMILAR FUNCTIONS BUT DIFFERENT ORIGINS.

THOSE STRUCTURES WHICH CAN NO LONGER BE CONSIDERED FUNCTIONAL SUCH AS THE APPENDIX.

A STUDY OF THE DISTRIBUTIONS AND ORIGINS OF PLANTS AND ANIMALS.

A THEORY WHICH SUGGESTS THAT THE ORIGIN OF A SPECIES IS APPROXIMATELY AT THE CENTER OF ITS DISTRIBUTION.

NO TWO SPECIES CAN COMPETE INDEFINITELY.

UNRELATED ORGANISMS WHICH HAVE SIMILAR STRUCTURES SUCH AS THE EYE OF A SQUID AND THE HUMAN EYE.

WHAT AN ORGANISM DOES IN THE ENVIRONMENT, SOMEWHAT LIKE THE ORGANISMS JOB.

THE PROCESS OF A SINGLE ANCESTOR EVOLVING INTO A VARIETY OF FORMS OR SPECIES WHICH OCCUPY DIFFERENT HABITATS AND NICHES.

THE PROCESS OF EVOLUTION WHICH LEADS TO WIDELY DIFFERING ADAPTATIONS TO A VARIETY OF ENVIRONMENTAL CONDITIONS.

A METHOD TO DATE ROCKS AND FOSSILS.

A METHOD USED TO DATE ROCKS AND FOSSILS.

CHAPTER 35

BEHAVIOR

BEFORE READING ACTIVITIES

1. Chapter 35 presents a wide range of topics associated with behavior. It is important for you to get an overview of what will be discussed before you begin to read. An important reason for getting an overview before reading is that what you understand about new information depends upon what information you already have about the subject. This means that the knowledge you have now influences how you will interpret what you read. To get an overview of what concepts will be discussed, turn to the outline on page 724 of the text. Read the outline and answer the following questions.

    1. How many behavior related topics will be discussed in this chapter?

    2. Which topics do you already know something about?

    3. What do you know about each topic? For example, what do you know about behavior? How do you think this information is related to biological concerns?

    4. What do you know about learning? How do you think this information is related to biological concerns?

    5. Write down what you already know about a topic and how you think this information is related to biological concerns.

2. Read the Learning Objectives on page 724. You should notice the following points as you read the objectives:

    1. Some of these objectives ask you to complete relatively simple activities: summarize, cite examples, define, and present a concept.

    2. Some of these objectives ask you to complete much more complex tasks: support the theses, compare, distinguish between, postulate, and the task in objective 9, given a description, identify...

    3. Read through the vocabulary terms. Note that many of the terms are ones you have learned earlier in your reading or contain word parts you already know (such as endo- or exo-).

4. Read the introduction (note that it is an example rather than an overview of the chapter), the summary (it is quite long) and use your knowledge of the vocabulary to take the Post-Test on pages 750 and 751.

DURING READING ACTIVITIES

1. As you read, begin to prepare your answers for each of the objectives at the beginning of the chapter.

   1. Objective 11 asks you to DISTINGUISH BETWEEN two things (home range and territory). Distinguish between means to EXPLAIN THE DIFFERENCE between things. (COMPARE means the same thing.) Begin a chart which shows how home range and territory are different. Your chart should have two columns. (Columns are vertical.) To determine how many rows you should have, you will need to decide what characteristics you should include. (Rows are horizontal.) You should find the information you need on pages 739 and 740.

   2. Objectives 6 and 13 ask you to COMPARE two things. When you compare you need to give BOTH LIKENESSES AND DIFFERENCES.

   3. Objectives 8 and 10 ask you to "speculate" and "postulate". In order to SPECULATE or POSTULATE you will need to MAKE INFERENCES from information that is presented in the text. Since there is not a definite, correct answer, you need to use terms of qualification (similar to those the authors used in Chapter 34). Some of the words you can use include: may, might, likely, it is suggested, it is possible that.

2. Be sure to circle new vocabulary and underline the definitions. Check to see if there are vocabulary cards for terms new to you. If there are not, make your own cards.

AFTER READING ACTIVITIES

1. Review what you have read. Check your textbook marking to be sure it is accurate.

2. Go through the objectives and write or say aloud your answers.

3. Answer the following questions.

   1. The study of BEHAVIOR under natural conditions is termed _______ which tends to be _______________ and helps to maintain homeostasis.

   2. _____________ BEHAVIOR is exemplified by ______________ in plants and by __________________ in animals.

   3. List three types of TROPISMS and three types of TAXES and explain briefly what they are.

4. A behavioral rhythm which is based on the cycle of the moon is termed ________________ ______________, those based on about a 24 hour period are _________________ _______________ and BIOLOGICAL _____________ can be based on internal mechanisms called _________________ MECHANISMS or external mechanisms called ______________ MECHANISMS.

5. _______________ BEHAVIOR is under genetic control whereas ___________ BEHAVIOR requires a previous experience.

6. List the requirements for SOCIETAL BEHAVIOR.

7. __________________ among animals requires the transmission of signals between individuals. A type of chemical COMMUNICATION is termed a ________________.

8. The structure in the society which tends to suppress aggressive behavior is termed a _______________ ______________. A member of a species lives in an area which is called a _______________ _________, in certain areas an animal may defend its __________________.

9. ____________________ BEHAVIOR assures that the appropriate sexes and species mate. A _____________________ is the behavior which holds members of the opposite sex in a mutual stable relationship.

10. The human society is the most complex of all animal behaviors. List the requisites of the human society.

Chapter 35 - Answers

1. ethology, adaptive
2. simple, tropisms, taxes
3. Phototropism--response to light, geogotropism--response to gravity, thigmotropism--response to touch, geotaxis--response to gravity, phototaxic--response to light, chemotaxic--response to chemicals)
4. lunar cycle, circadadian rhythms, clocks, endogeneous, exogenous
5. innate, learned
6. SOCIAL BEHAVIOR is a CONSPECIFIC INTERACTION where members frequently put the society above self. There must be a form of COMMUNICATION, members are divided into DIVISIONS OF LABOR and PARENTAL PROVISIONING is provided the young.)
7. communication, phermomone
8. dominance hierarchy, home range
9. courtship, pair bond
10. The human society requires a LANGUAGE, CULTURE and TOOL USAGE.

ETHOLOGY

BEHAVIOR

SIMPLE BEHAVIOR

ADAPTIVE

TROPISISMS

TWIDDLES

GEOTROPISM

PHOTOTROPISMS

TAXES

THIGMOTROPISM

PHOTOTAXIS

GEOTAXIS

THE RESPONSES OF AN ORGANISM TO SIGNALS IN THE ENVIORNMENT

THE STUDY OF BEHAVIORAL ADAPTATIONS IN THE NATURAL ENVIRONMENT

HAVING A VALUE IN THE ENVIRONMENT

SIMPLE ONE FUNCTION MOVEMENTS OR REACTIONS

RANDOM MOTIONS OF BACTERIA IN CLOCKWISE AND COUNTERCLOCKWISE DIRECTIONS, SIMPLE BEHAVIOR

GROWTH RESPONSES IN PLANTS TOWARD OR AWAY FROM A STIMULUS

RESPONSES TO LIGHT

RESPONSE TO GRAVITY

RESPONSE TO A SOLID OBJECT OR TOUCH

ORIENTATIONAL BEHAVIORS OF ANIMALS

RESPONSE TO GRAVITY

RESPONSE TO LIGHT

THIGMOTAXIC

CHEMOTAXIC

CIRCADIAN RHYTHMS

LUNAR CYCLE

NOCTURNAL

DIURNAL

ENDOGENOUS

CREPUSCULAR

INSTINCT OR INNATE

EXOGENOUS

SIGNED STIMULI OR RELEASERS

LEARNED BEHAVIOR

RESPONSE TO CHEMICALS	RESPONSE TO SOLID OBJECT OR TOUCH
BEHAVIORS WHICH ARE SYNCHRONIZED WITH THE CYCLE OF THE MOON	PERIODS OF BEHAVIORAL ACTIVITY BASED ON APPROXIMATELY 24 HOUR CYCLE
ACTIVE DURING THE DAY	ACTIVE DURING THE NIGHT
ACTIVE IN THE DAWN AT DUSK OR BOTH	BIOLOGICAL CLOCKS WHICH ARE REGULATED FROM WITHIN THE ORGANISM
BIOLOGICAL CLOCKS WHICH ARE REGULATED BY EXTERNAL OR ENVIRONMENTAL STIMULI	BEHAVIOR IS UNDER GENETIC CONTROL
BEHAVIOR IS DEVELOPED AS A RESULT OF EXPERIENCE	USUALLY SERVE AS TRIGGERS FOR FIXED PATTERNS OF BEHAVIOR SUCH AS ALARM BEHAVIOR

IMPRINTING

INSIGHT LEARNING

ZUGUNRUHE

MIGRATION

CONSPECIFIC

SOCIAL BEHAVIOR

PHEROMONE

COMMUNICATION

HOME RANGE

DOMINANCE HIERARCHY

TERRITORIALITY

TERRITORY

THE ABILITY TO REMEMBER A PAST EXPERIENCE OR STIMULI AND BE ABLE TO RESPOND TO A SIMILAR STIMULI IN A DIFFERENT SITUATION AND ADAPT THE RESPONSE TO THE NEW SITUATION

WHEN A YOUNG ANIMAL FORMS A STRONG BOND TO AN INDIVIDUAL, USUALLY A PARNT, SHORTLY AFTER BIRTH

A DIRECTIONAL AND TIMED MOVEMENT OF A POPULATION

MIGRATORY RESTLESSNESS

INTERACTIONS (COMMUNICATION, DIVISION OF LABOR, AND PARENTAL CARE) ADAPTIVE TO ALL MEMBERS WITHIN THE SOCIETY

AMOUNG MEMBERS OF THE SAME SPECIES

WHEN ONE ANIMAL PERFORMS AN ACT THAT CHANGES THE BEHAVIOR OF ANOTHER ANIMAL

THE RELEASE OF A CHEMICAL BY ONE INDIVIDUAL WHICH CAUSES ANOTHER INDIVIDUAL TO RESPOND TO THE SIGNAL, USUALLY ASSOCIATED WITH THE COURTSHIP

AN ARRANGEMENT OF STATUS WITHIN A SOCIETY WHICH REGULATES AGGRESSIVE BEHAVIOR

THE AREA WHICH AN ORGANISM LIVES IN

THE AREA, USUALLY WITHIN THE HOME RANGE, WHICH AN ANIMAL WILL DEFEND

THE TENDENCY TO DEFEND A TERRITORY

PAIR BOND

COURTSHIP

ROYAL JELLY

PARENTAL CARE

ROUND DANCE

DANCE

HYGIENIC BEHAVIOR

WAGGLE DANCE

ALTRUISTIC BEHAVIOR

CULTURE

COEFFICIENT OF RELATEDNESS

KIN SELECTION

THE BEHAVIOR OF MATE SELECTION BY MEMBERS OF THE SAME SPECIES	A STABLE, USUALLY PERMANENT, PAIRING OF TWO MEMBERS OF THE OPPOSITE SEX WITHIN A SPECIES
PROVIDING FOR, USUALLY FOOD, FOR THEIR YOUNG	A SUBSTANCE SECRETED BY NURSE BEES ESSENTIAL TO THE GROWTH OF ALL BEES
COMMUNICATION AMONG BEES	THE DANCE PERFORMED BY BEES IF A FOOD SOURCE IS NEARBY
THE DANCE PERFORMED BY BEES IF A FOOD SOURCE IS DISTANT, COMMUNICATES DISTANCE AND DIRECTION	THE SPECIFIC BEHAVIOR IN BROWN BEES IN WHICH THE CELLS OF DEAD LARVA ARE CLEANED REGULARLY
THE COMPLEX BEHAVIOR SHOWN IN HUMAN SOCIETIES TYPIFIED BY COMMUNICATION BY A LANGUAGE AND TOOL USE	AN INDIVIDUAL WORKING IN BEHALF OF THE SOCIETY RATHER THAN ITSELF
THE PROCESS WHEREBY THE GENES OF THE DOMINANTS IN THE HIERARCHY ARE PAIRED	A FORMULA TO EXPRESS KINSHIP

SOCIOBIOLOGY

THE STUDY OF BEHAVIOR THROUGH
NATURAL SELECTION

CHAPTER 36

ECOLOGY: THE BASIC PRINCIPLES

BEFORE READING ACTIVITIES

1. This chapter is the first of three chapters that discuss the same general topic: ecology. In chapter 36 the basic principles underlying information discussed in chapters 37 and 38 are presented. This means that you need to understand these principles if you are to understand the more complex information in the later chapters.

2. To get an overview of the this chapter, turn to page 753 and read the outline. Answer the following questions.

   1. What are the most important concepts discussed?

      1.   3.

      2.   4.

      3.   6,

   2. What do you already know about these concepts? Do you have any idea what limiting factors, competition, or predation are?

   3. What do you know about nutritional life-styles? Do you think your information is similar to what will be discussed in the text?

   4. What are the dynamics of ecosystems?

      1.   3.

      2.   4.

      3.   6.

   5. What do you think "succession" is?

3. Read the Learning Objectives. Notice the following points as you read:
   1. Many of the objectives ask you to do MORE THAN ONE thing. For example, Objective 1 asks you to define, distinguish among, and give examples. Objective 2 asks you to distinguish between...and relate them to...It is important to read all of the directions carefully and to underline the key terms so you will not forget to complete a part of the required activity. (This is especially true when you must answer essay questions on an exam.)

   2. Objective 5 asks you to draw a food chain. It suggests that this can (and should) be done for a variety of organisms. This is a very complex activity requiring you to use the principles given in the text, the example given in Figure 36-13 and your own ability to make inferences and apply all of this information to new situations. You should practice making food chains so you will be ready to create a new one if you are asked to do so on an exam.

4. Read through the vocabulary terms. Some information you should be aware of includes the following:

   1. There are several pairs of words which have common word bases but different prefixes. You need to know what the base word means and how the prefixes are different. Examples include intraspecific and interspecific, autotroph and heterotroph, photosynthetic and chemosynthetic.

   2. There are a number of words used in this chapter that you learned earlier, such as symbionts and trophic. (Refer to your vocabulary cards if you have forgotten these words).

   3. There are words which seem to be common words, but which have some specific and different meanings in this context. Find the specific meanings.

5. Read the introduction and summary, and take the Post-Test on page 775.

DURING READING ACTIVITIES

1. As you read, begin to prepare your answers for each of the objectives. When you finish reading a section, review it (by rereading your textbook marking and thinking about what all those words really mean), and then write your answer for the Learning Objectives related to it.

2. Continue to circle the new vocabulary and underline the definitions of the terms.

3. Read the verbal portion that goes with each figure carefully. Write your summary of each figure in the margin.

AFTER READING ACTIVITIES

1. Review what you read. Check your textbook marking to be sure it is accurate.

2. Reread the chapter outline and recite what you know about each topic.

3. Test yourself on the vocabulary terms. Practice reciting or writing the definitions for those you don't know in groups of 5-7 words.

4. Answer the following questions.

   1. Briefly describe the units of an ECOSYSTEM

   2. Name the three major roles which organisms play in the ECOSYSTEM.

   3. The ____________________ is the place where an organism lives. The ____________________ is the role which the organism plays in the COMMUNITY.

4. In nature POPULATIONS are not allowed to reach their full numbers or ____________________. A variety of ____________________ are dependent on numbers within the population and are called ______________ ______________ ___________ _____________. Other factors which contribute to ___________________ ______________ are interspecific and intraspecific _____________________ ___________ are interspecific and intraspecific _______________.
On some occasions one population may be living entirely at the expense of another individual, this is called _______________.

5. Organisms which live exclusively on meat are called ___________; those which feed on plants only are termed _______________, and those who have a mixed diet are called __________________.

6. Distinguish between a FOOD CHAIN, FOOD WEB and TROPHIC LEVELS.

7. Briefly state how raw resources remain in the environment and indicate the variety of materials which are recycled in the environment.

8. Briefly describe how COMMUNITIES develop.

Chapter 36 - Questions

1. An ECOSYSTEM is made up of a variety of COMMUNITIES including a consideration of NON-BIOTIC factors. COMMUNITIES are made up of a consideration of NON-BIOTIC factors. COMMUNITIES are made up of a variety of POPULATIONS which exist in a certain area, in turn POPULATIONS are groups of individuals of the same SPECIES.

2. Organisms can be PRODUCERS in which case they provide the bases of the FOOD PYRAMID. CONSUMERS are those organisms which feed on PRODUCERS. DECOMPOSERS are responsible for recycling the nutrients back into the ECOSYSTEM.

3. habitat, ecological niche

4. biotic potential, limiting factors, population dependent limiting factors, environmental resistance, competition, predation.

5. carnivores, herbivores, omnivores

6. A FOOD CHAIN is the linear progression of which organism feeds on which organism; whereas a FOOD WEB is a complex interaction of which organisms interconnect in their feeding habits. TROPHIC LEVELS refer to the transfer of energy between levels of a food chain.

7. According to the LAWS OF THERMODYNAMICS raw materials are never lost. As a result organic materials which break down into their inorganic components can also be built back up into organic compounds. Thus, NITROGEN, SULFUR, POTASSIUM, PHOSPHORUS, IRON to name a few.)

8. ECOLOGICAL SUCCESSION is the process whereby a COMMUNITY within the ECOSYSTEM undergoes a progression of changes in its biotic elements. Typically a new or disturbed area will not be colonized by PIONEER species and subsequently other organisms will immigrate into the area end in a stable CLIMAX COMMUNITY.)

POPULATION

ECOLOGY

ECOSYSTEM

COMMUNITY

HABITAT

ECOSPHERE

BIOTIC POTENTIAL

ECOLOGICAL NICHE

NONBIOTIC FACTORS

ENVIROMENTAL RESISTANCE

LIMITING FACTOR

BIOTIC FACTORS

THE SUBDISCIPLINE OF BIOLOGY WHICH STUDIES THE INTERACTIONS BETWEEN AND AMONG ORGANISMS AND THEIR ENVIRONMENT

A GROUP OF INTERBREEDING INDIVIDUALS, A SPECIES, WHICH OCCUPY A PARTICULAR AREA

ALL THE POPULATIONS WHICH OCCUPY A SIMILAR AREA AT A GIVEN TIME

A COMMUNITY OF ORGANISMS IN RELATION TO THE NONLIVING ENVIRONMENT

THE PLANET EARTH AND ALL THE ORGANISMS IN IT

THE PLACE WHERE THE ORGANISM LIVES

IS THE ROLE OF THE SPECIES WHERE IT LIVES, HOW IT FUNCTIONS IN THE COMMUNITY

THE CAPACITY, UNCHECKED BY ANY FACTORS, FOR A POPULATION TO INCREASE ITS NUMBERS

A VARIETY OF FACTORS WHICH INFLUENCE THE BIOTIC POTENTIAL OF A POPULATION

ANY FACTOR WHICH DOES NOT INCLUDE SOME ORGANISM INTERACTION, SUCH AS CLIMATE; CLIMATE IS AN EXAMPLE

ANY FACTOR WHICH RELATES TO ORGANISMS AND THEIR INTERACTION

THE FACTOR WHICH IS IN THE SHORTEST SUPPLY OR WHICH INFLUENCES A POPULATION IN SOME WAY

POPULATION DEPENDENT LIMITING FACTORS

POPULATION INDEPENDENT LIMITING FACTOR

INTERSPECIFIC COMPETITION

INTRASPECIFIC COMPETION

APOSEMATIC

PREDATION

HETEROTROPHS

AUTOTROPHS

CARNIVORES

HERBIVORES

SYMBIONT

OMNIVORE

A FACTOR WHICH IS NOT REGULATED BY NUMBERS OF INDIVIDUALS

A FACTOR WHICH IS REGULATED BY NUMBERS OF INDIVIDUALS

COMPETITION WITHIN A SPECIES, BETWEEN INDIVIDUALS WITHIN A POPULATION, WHICH COMPETE FOR FOOD OR SHELTER OR A PLACE TO LIVE

TWO DIFFERENT SPECIES COMPETING FOR THE SAME RESOURCES

ONE POPULATION LIVING AT THE EXPENSE OF ANOTHER POPULATION CALLED THE PREY

A COLORATION PATTERN WHICH WARNS PREDATORS IN SOME WAY

ORGANISMS WHICH ARE CAPABLE OF MAKING THEIR OWN FOOD SUCH AS PLANTS SOMETIMES REFERRED TO AS PRODUCERS, CAN BE PHOTOSYNTHETIC WHICH REQUIRE SUNLIGHT OR CHEMOSYNTHESIZERS WHICH OXIDIZE INORGANIC SUBSTANCES

ORGANISMS WHICH CANNOT MAKE THEIR OWN FOOD AND MUST RELY ON OTHER SOURCES OF NUTRITION SUCH AS ANIMALS SOMETIMES REFERRED TO AS CONSUMERS

FEED EXCLUSIVELY ON PLANT LIFE

FEED EXCLUSIVELY ON ANIMAL LIFE

FEED ON BOTH PLANT AND ANIMAL LIFE

MEANS TO LIVE TOGETHER IN SOME SORT OF A RELATIONSHIP

SAPROBIC OR SAPROTROPHS

DETRITUS FEEDER OR SCAVENGER

FOOD CHAIN

CLEANING SYMBIOSIS

PRIMARY CONSUMER

TROPHIC LEVEL

FOOD WEB

SECONDARY CONSUMER

BIOMASS

PYRAMID OF NUMBERS

NET PRODUCTION

GROSS PRODUCTION

FEED ON DECOMPOSING ORGANIC MATERIALS

ORGANISMS WHICH ABSORB THEIR NUTRITION

THE PHENOMENON OF ONE ORGANISM CLEANING ANOTHER AND BOTH BENEFIT FROM THE RELATIONSHIP

A SERIAL ARRANGEMENT OF WHICH ORGANISM FEEDS ON WHICH ORGANISM

THE LEVELS OF PRODUCTION OR COMSUMPTION IN A FOOD PYRIMID

THE FIRST TROPHIC LEVEL WHICH FEEDS DIRECTLY ON THE PRODUCER

THE NEXT TROPHIC LEVEL BEYOND PRIMARY CONSUMER

FOOD CHAINS OF A COMMUNITY WHICH INTERCONNECT IN A VARIETY OF WAYS

A COUNT OF THE NUMBERS AT EACH TROPHIC LEVEL

THE LIVE WEIGHT OF ORGANIC MATERIAL

THE RATE OF TOTAL ENERGY STORAGE

GROSS PRODUCTION MINUS THE ENERGY USED IN THE FROM WITHIN THEY SYSTEM

BIOCHEMICAL CYCLES

COMMUNITY PRODUCTIVITY

PIONEER COMMUNITY

ECOLOGICAL SUCCESSION

PRIMARY SUCESSION

CLIMAX COMMUNITY

SECONDARY SUCESSION

THE RATE AT WHICH ENERGY IS STORED IN THE FORM OF NEW BIOMASS

THE RECYCLING OF NUTRIENTS IN THE ENVIRONMENT

A PROGRESSIVE SERIES OF CHANGES IN THE KINDS AND NUMBERS OF ANIMALS IN A GIVEN COMMUNITY

THE FIRST ORGANISMS TO APPEAR IN A NEW OR DISTURBED AREA

THE END POINT IN THE DEVELOPMENT OF A COMMUNITY, ONE WHICH HAS REACHED MATURITY

THE EARLY DEVELOPMENT WITHIN A COMMUNITY, THE INFLUX OF PIONEERS

BEGINNING WITH A PRE-EXISTING COMMUNITY AND IN SOME WAY HAVING ADDITIONAL NEW DEVELOPMENT TAKE PLACE

CHAPTER 37

ECOLOGY: THE MAJOR COMMUNITIES

BEFORE READING ACTIVITIES

1. Before reading chapter 37, review chapter 36. Remember that the concepts presented in chapter 36 will help you understand chapter 37. To review, use the Learning Objectives on page 753. As you read each objective, write your answer or say it aloud. Compare your answer with the one you wrote while you were reading the chapter.

2. To get an overview of the information to be presented in chapter 37, turn to page 777 and read the outline. Answer the following questions.

    1. What are the major ecological communities?

        1.

        2.

    2. What are the terrestrial life zones?

        1. 3.

        2. 4.

    3. What are the Aquatic habitats?

        1.

        2.

    4. What are the different types of life in the sea?

        1. 3.

        2. 4.

3. Read the Learning Objectives on page 777. Although there are only 5 Learning Objectives, some of them are quite complicated because they ask you to do more than one thing or they ask you to do the same thing for several topics.

    1. Look at Learning Objective 2. It asks you to describe AT LEAST THREE things which differ SIGNIFICANTLY. This is a complicated task to fulfill for two reasons. First, the "at least" suggests that 3 is a minimum and that 4 or 5 would probably be better. Second, "SIGNIFICANTLY" means that you must use IMPORTANT differences in your description. To use important differences you will need to think through all the information you have and evaluate what is important and what is not important. EVALUATE means that you cannot just use any information you know; you must only use the most important information.

    2. Look at Learning Objective 3. It, too, is a complicated task with two parts and each of those 2 parts have subsections. You must first describe some marine communities and you must use some specific information in the description. Then you must describe more than one community and you need to use information about 4 specific factors to describe each community. One way to prepare yourself to find this information while you are reading is to set up a table. The table can be set up like the one that follows.

DESCRIPTION OF MARINE COMMUNITIES

Community	Food-Web	Zones of Productivity	Constraints	Significance
estuarine				
intertidal				
abyssal				
?				

4. Read through the vocabulary cards and start now to learn the terms that are unfamiliar.

5. Read the introduction, summary and take the Post-Test.

DURING READING ACTIVITIES

1. As you read, look for the information you need to complete each of the objectives. Write the information necessary to fulfill each objective.

2. Read the chapter from one major heading to another. As you complete a section, reread your textbook marking and correct any inadequate marking.

3. Read the Figures carefully. As you read them, write a summary of what each figure means and how it relates to the information given in the text.

AFTER READING ACTIVITIES

1. Review what you have read in the entire chapter. Go over your textbook marking.

2. Turn to the chapter outline on page 777 and test yourself. Write the information related to each major heading and then check to see if you have included the most important points.

3. Review the vocabulary.

4. Answer the following questions.

    1. List the eight BIOMES discussed in the text and characterize each one.

2. List the three major aquatic COMMUNITIES and characterize each one

3. List the major habitats of both fresh water and marine systems and characterize each one.

4. Describe the process of EUTROPHICATION.

Chapter 37 - Answers

1. A direct listing out of either the text or vocabulary list.

2. A direct listing out of the text, vocabulary or summary.

3. A direct listing out of the text, vocabulary or summary.

4. Aquatic ecosystems usually start out by being clear water with very little nutrients and few species. As species invade the body of water it accumulates more nutrients ending in the EUTROPHIC condition. This process can be speeded up drastically depending on the influence of humans.

ECOTONES

BIOMES

PERMAFROST

TUNDRA

TEMPERATE DECIDUOUS FOREST

TAIGA

DESERT

TEMPERATE GRASSLAND

TROPICAL RAIN FOREST

CHAPPARAL

STRANGLER TREE

EPIPHYTE

EASILY DIFFERENTIATED COMMUNITY UNITS BASED ON THE BIOTA WHICH LIVES THERE

AN AREA BETWEEN BIOMES WHICH HAS AN INTERGRADATION OF BIOTA FROM BOTH BIOMES

THE CIRCUMPOLAR BIOME PERMANENTLY COVERED BY ICE

THE LAYER OF FROZEN GROUND IN THE TUNDRA

THE CIRCUMPOLAR BIOME BELOW THE TUNDRA WHICH IS DOMINATED BY CONIFERS AND A DEEP BED OF SPRUCE AND PINE NEEDLES

THE BIOME LOCATED IN THE NORTHEASTERN AND MIDDLE EASTERN U.S., TYPICALLY OF HARDWOOD TREES AND A CLAY AND HUMUS LAYER BENEATH THEM

THE BIOME LOCATED IN THE MIDDLE WESTERN U.S., TYPICALLY SHORT OR TALL GRASSES WITH EXCELLENT SOIL QUALITIES

THE BIOME CHARACTERIZED BY TEMPERATURE EXTREMES AND SANDY SOIL WITH LITTLE HUMUS

A SPECIALIZED KIND OF DESERT BIOME CHARACTERIZED BY SHRUBS AND A FEW TREES AND DISTINCTIVE ABUNDENT RAINFALL DURING A SHORT SEASON

THE BIOME CHARACTERIZED BY MANY STORIES OR LEVELS OF VEGETATION, THE TEMPERATURE REMAINS CONSTANT WITH HIGH RAINFALL

PLANTS WHICH LIVE ON OTHER PLANTS

A PLANT WHICH IS ADAPTED TO CLIMB AND ATTACH TO A HOST TREE

STENOHALINE

SAVANNAH (=VELD)

LITTORAL ZONE

EURYHALINE

LIMNETIC ZONE

NEUSTON

COMPENSATION POINT

PROFUNDAL ZONE

THERMOCLINE

THERMAL STRATIFICATION

BLOOMS

FALL TURNOVER

THE BIOME WHICH IS A TROPICAL GRASSLAND

NARROWLY TOLERANT TO SALINITY

WIDELY TOLERANT TO SALINITY

THE ZONE OF THE SHORE SURROUNDING A BODY OF WATER

ORGANISMS WHICH LIVE AT THE AIR-WATER INTERFACE ON THE SURFACE

THE OPEN DEEPER PART OF THE BODY OF WATER

THE AREA WHICH DOES NOT RECEIVE ENOUGH LIGHT TO SUPPORT PLANT LIFE

THE PRODUCTS OF PHOTOSYNTHESIS ARE OFFSET BY THE INTAKE OF CONSUMERS WITH NO NET GAIN

THE TEMPERATURE LAYERING IN A BODY OF WATER WHERE THE WARMEST WATER IS ON THE TOP

THE AREA IN THE WATER COLUMN OF ABRUPT TEMPERATURE DIFFERENCES

THE COMPLETE MIXING OF LAKE WATERS DUE TO WINDS AND TEMPERATURE CHANGES

THE RAPID DEVELOPMENT OF ALGAE GROWTH ON THE SURFACE OF THE BODY OF WATER

EUTROPHIC

OLIGOTROPHIC

PLANKTON

CULTURAL EUTROPHICATION

BENTHOS

NEKTON

ESTUARY

UPWELLING

SUBTIDAL ZONE

INTERTIDAL ZONE

DEEP WATERS

NERITIC ZONE (=CONTINENTAL SHELF)

CLEAN WATER LAKES WITH LITTLE SPECIES DIVERSITY, EARLY IN THEIR SUCCESSION

LAKES RICH WITH NUTRIENTS AND HIGH SPECIES DIVERSITY

THE PROCESS OF BECOMING EUTROPHIC BECAUSE OF HUMAN PRACTICES OF PROVIDING WATER WITH EXCESS NITRATES AND SEPTIC DRAINAGE

MICROSCOPIC ORGANISMS WHICH ARE THE BASE OF AQUATIC FOOD CHAINS

ORGANISMS WHICH CAN MOVE BY MEANS OF THEIR OWN POWER

BOTTOM DWELLING ORGANISMS

DEEP CURRENTS WHICH BRING NUTRIENTS TO THE SURFACE

COSTAL WATER WHICH HAS ACCESS TO THE SEA AND FEED BY FRESH WATER

THE AREA BETWEEN LOW AND HIGH TIDE

BELOW THE LOWEST TIDE, PROTECTED FROM WAVE ACTION AND IS IN SHALLOW ENOUGH WATER FOR PHOTOSYNTHESIS TO TAKE PLACE

WATERS OF THE SHELF LESS THAN 200 FEET DEEP

WATER MORE THAN A MILE DEEP

CHAPTER 38

HUMAN ECOLOGY

BEFORE READING ACTIVITIES

1. Before reading chapter 38, review chapters 36 and 37. All three of these chapters discuss aspects of the same subject: ecology, so it is important that you recall the information presented in previous chapters before you read this chapter. To review, go back to the chapter outline or the Learning Objectives on page 777 to write your responses or say them aloud.

2. To get an overview of the information presented in chapter 38, turn to page 797 and read the chapter outline. Then answer the following questions.
   1. What are the human ecological problems which will be discussed?

      1.   6.
      2.   7.
      3.   8.
      4.   9.
      5.   10.
           11.

   2. Which problems seem to be more complex? (Judging by the amount of topics needed to discuss them.)

      1.   3.
      2.   4.

   3. What do you already know about each of these topics? Do you already have an opinion about what should or should not be done about these problems? Do you believe that these problems will have any effect on your life? If so, what are the possible effects?

3. Read the Learning Objectives listed below. As you read them, underline the key words which tell you what to do.

KEY WORD	LEARNING OBJECTIVES	INFORMATION FOUND
1. ________	1. Review the (1) development and (2) impact of modern human life-styles upon the ecosystems of of the earth.	________
2. ________	2. Contrast an agricultural community with the natural community it replaces, relating the differences to ecological instability.	________
3. ________	3. Describe two problems associated with pesticide use and two problems specifically associated with the use of persistent pesticides (chlorinated hydrocarbons)	________
4. ________	4. Summarize the sources of water pollution and describe the effects of dumping organic wastes into a stream.	________
5. ________	5. Discuss the (1) principal ecological effects and (2) climatic implications of air pollution.	________
6. ________	6. Describe the principal methods of solid-waste disposal and discuss the relevance of recycling to waste disposal.	________
7. ________	7. Discuss (1) nuclear power and (2) solar power as energy options.	________
8. ________	8. Summarize the process of extinction, (1) listing factors that contribute to the decline and extinction of endangered species, and providing an example of each.	________
9. ________	9. Relate overpopulation to specific environmental problems and explain how humans can temporarily expand the carrying capacity of their habitat.	________

1. Write the key word(s) in the column Key Words. As you do this, you will notice that in many of these objectives you are asked to do more than one thing.

2. If you are asked to do something for more than one subject, number the subjects you must do something for. For example, in objective 1 you are asked to review the (1) development and (2) impact of .... In objective 7, you are asked to discuss (1) nuclear power and (2) solar power.

3. After you have determined what you need to do, locate the places in the text where the information will be found. Write the page numbers in the column "Information Found."

4. Read through the vocabulary cards and determine which terms you already know and which terms you need to study.

5. Read the introduction and summary sections of the chapter. Take the Post-Test on page 820.

DURING READING ACTIVITIES

1. Read the chapter from one major heading to another. As you finish reading a section, reread your textbook marking and think about what you have learned.

2. Look for the information you need to complete each of the objectives. Write the appropriate information on a separate sheet of paper.

3. Carefully look at the pictorial content of the figures. Try to decide what a figure illustrates without reading the verbal portion. After you have "written" your own explanation, read the explanation in the text and compare your interpretation with the authors'. Remember to write down a summarization of what the figure means.

    1. For example, you might summarize figure 38-8 by writing that it, "shows a sequence of events that occur when organic wastes are in the water."

AFTER READING ACTIVITIES

1. Review the information in the entire chapter. Correct any errors you made while marking your textbook.

2. Review the vocabulary for this chapter and chapters 36 and 37.

3. Answer the following questions.

    1. An agricultural technique which simplifies the community is termed ____________________. This is the development and planting of a single crop, like a Christmas Tree Farm which could be called ____________________.

    2. Identify the four characteristics which set apart agriculture practices from undisturbed environments.

    3. The first generation pesticides were ____________________ which were quickly replaced by organic compounds in three groups ____________________ ____________________. ____________________ ____________________, and ____________________.

    4. What are the major reasons that DDT and other organic compounds are a problem?

5. The two major sources of water pollution are ____________ ____________ and ________________ ________________.

6. The four most common pollutants of air are: ____________ ____________, ________________ ________________, ________________ ________________, and ________________.

7. Two other forms of pollution are ________________ which is the result of nuclear weapons testing and ____________ ____________ ____________ which U.S. citizens produce in the amount of 3.6 kilograms per day.

8. The two options for energy are ________________ and ____________ power.

9. Why do organisms become extinct?

Chapter 38 - Answers

1. monoculture, silvaculture

2. Agricultural communities are: (1) unstable, (2) monocultures which attract host-pest interactions, (3) chemical cycles are incomplete and interrupted and (4) competition between species is disturbed and reduced.

3. inorganic chemicals, chlorinated hydrocarbons, organophosphates, carbamates.

4. For the most part, organic pesticides are persistent for long periods of time. They do not remain in the areas where they are sprayed. Also, organic compounds are not easily metabolized and tend to accumulate and multiply in fat tissues over time.

5. industrial wastes, municipal sewage

6. carbon monoxide, sulfur oxide, nitrogen oxides, hydrocarbons.

7. radioactive, solid waste disposal

8. nuclear, solar

9. Answer is directly in text.

SILVICULTURE

MONOCULTURE

SECOND GENERATION PESTICIDES

FIRST GENERATION PESTICIDES

PERSISTANT PESTICIDES

BIODEGRADABLE

POLLUTION

BIOLOGICAL MAGNIFICATION

ATMOSPHERIC INVERSION

SINK

SANITARY LANDFILL

GREENHOUSE EFFECT

THE DEVELOPMENT AND CULTURING OF A SINGLE SPECIES SUCH AS GROWING WHEAT

THE DEVELOPMENT AND CULTURING OF A SINGLE SPECIES OF TREE SUCH AS A CHRISTMAS TREE FARM

INORGANIC CHEMICALS TO REPEL OR KILL INSECTS

SYNTHETIC ORGANIC COMPOUNDS TO KILL INSECTS

RAPID DECOMPOSITION

PESTICIDES WHICH DO NOT READILY DEGRADE

THE PROCESS WHEREBY PESTICIDES INCREASE IN THEIR ACCUMULATION ACCORDING TO HOW MANY ORGANISMS ARE IN THE FOOD CHAIN

THE REDUCTION IN THE QUALITY OF THE ENVIRONMENT

A DEPOSITING GROUND FOR UNDESIRABLE MATERIALS

ATMOSPHERIC CONDITIONS WHICH CAUSE AIR TO ACCUMULATE POLLUTANTS, WARM AIR ON TOP OF COLDER AIR

HOLDING IN HEAT, THUS THE TEMPERATURES RISE

A PLACE DESIGNED TO ACCEPT DISPOSAL OF SOLID WASTES

DESTRUCTION OF HABITAT

FISSION PLANTS

LOGARITHMIC PHASE

LAG PHASE

CARRYING CAPACITY

EQUILIBRIUM

THE SPLITTING OF ATOMS TO PRODUCE ENERGY

THE TAKING AWAY OR MODIFYING NATURAL ENVIRONMENTS

SLOW INCREASE IN POPULATION SIZE DURING THE EARLY PART OF INCREASING NUMBERS

THE PORTION OF A GROWTH CURVE CHARACTERIZED BY EXPONENTIAL GROWTH

THE PORTION OF A GROWTH CURVE WHERE THE NUMBERS STOP INCREASING

THE MAXIMUM POPULATION THAT AN ENVIRONMENT IS CAPABLE OF SUSTAINING